CANCER PROTEOMICS

CANCER DRUG DISCOVERY AND DEVELOPMENT

BEVERLY A. TEICHER, SERIES EDITOR

Checkpoint Responses in Cancer Therapy, edited by *Wei Dai*, 2008
Cancer Proteomics: *From Bench to Bedside*, edited by *Sayed S. Daoud*, 2008
Antiangiogenic Agents in Cancer Therapy, *Second Edition*, edited by *Beverly A. Teicher and Lee M. Ellis*, 2007
Apoptosis and Senescence in Cancer Chemotherapy and Radiotherapy, *Second Edition*, edited by *David A. Gerwitz, Shawn Edan Holtz, and Steven Grant*, 2007
Molecular Targeting in Oncology, edited by *Howard L. Kaufman, Scott Wadler, and Karen Antman*, 2007
In Vivo Imaging of Cancer Therapy, edited by *Anthony F. Shields and Patricia Price*, 2007
Transforming Growth Factor-β in Cancer Therapy, *Volume II: Cancer Treatment and Therapy*, edited by *Sonia Jakowlew*, 2007
Transforming Growth Factor-β in Cancer Therapy, *Volume 1: Basic and Clinical Biology*, edited by *Sonia Jakowlew*, 2007
Microtubule Targets in Cancer Therapy, edited by *Antonio T. Fojo*, 2007
Cytokines in the Genesis and Treatment of Cancer, edited by *Michael A. Caligiuri, Michael T. Lotze, and Frances R. Balkwill*, 2007
Regional Cancer Therapy, edited by *Peter M. Schlag and Ulrike Stein*, 2007
Gene Therapy for Cancer, edited by *Kelly K. Hunt, Stephan A. Vorburger, and Stephen G. Swisher*, 2007
Deoxynucleoside Analogs in Cancer Therapy, edited by *Godefridus J. Peters*, 2006
Cancer Drug Resistance, edited by *Beverly A. Teicher*, 2006
Histone Deacetylases: *Transcriptional Regulation and Other Cellular Functions*, edited by *Eric Verdin*, 2006
Immunotherapy of Cancer, edited by *Mary L. Disis*, 2006
Biomarkers in Breast Cancer: *Molecular Diagnostics for Predicting and Monitoring Therapeutic Effect*, edited by *Giampietro Gasparini and Daniel F. Hayes*, 2006
Protein Tyrosine Kinases: *From Inhibitors to Useful Drugs*, edited by *Doriana Fabbro and Frank McCormick*, 2005
Bone Metastasis: *Experimental and Clinical Therapeutics*, edited by *Gurmit Singh and Shafaat A. Rabbani*, 2005
The Oncogenomics Handbook, edited by *William J. LaRochelle and Richard A. Shimkets*, 2005
Camptothecins in Cancer Therapy, edited by *Thomas G. Burke and Val R. Adams*, 2005

Cancer Proteomics
From Bench to Bedside

Edited by

Sayed S. Daoud

Department of Pharmaceutical Sciences
Washington State University
Pullman, WA

Humana Press ✳ Totowa, New Jersey

© 2008 Humana Press Inc.
999 Riverview Drive, Suite 208
Totowa, New Jersey 07512
www.humanapress.com

All rights reserved. No part of this book may be reproduced, stored in a retrieval system, or transmitted in any form or by any means, electronic, mechanical, photocopying, microfilming, recording, or otherwise without written permission from the Publisher.

All articles, Comments, opinions, conclusions, or recommendations are those of the author(s), and do not neccessarily reflect the views of the publisher.

Due diligence has been taken by the publishers, editors, and authors of this book to assure the accuracy of the information published and to describe generally accepted practices. The contributors herein have carefully checked to ensure that the drug selections and dosages set forth in this text are accurate and in accord with the standards accepted at the time of publication. Notwithstanding, as new research, changes in government regulations, and knowledge from clinical experience relating to drug therapy and drug reactions constantly occurs, the reader is advised to check the product information provided by the manufacturer of each drug for any change in dosages or for additional warnings and contraindications. This is of utmost importance when the recommended drug herein is a new or infrequently used drug. It is the responsibility of the treating physician to determine dosages and treatment strategies for individual patients. Further it is the responsibility of the health care provider to ascertain the Food and Drug Administration status of each drug or device used in their clinical practice. The publisher, editors, and authors are not responsible for errors or omissions or for any consequences from the application of the information presented in this book and make no warranty, express or implied, with respect to the contents in this publication.

Production Editor: Rhukea Hussain
Cover design by Karen Schulz
Cover Illustration: *"The Power of Mass Spectrometry in Target Identification"* Derived from Fig. 4 of Chapter 2 by Kyunghee Lee, et al., and Fig. 3A of Chapter 5 by Christina M. Annunziata, et al.
This publication is printed on acid-free paper. ∞
ANSI Z39.48-1984 (American National Standards Institute)
Permanence of Paper for Printed Library Materials

For additional copies, pricing for bulk purchases, and/or information about other Humana titles, contact Humana at the above address or at any of the following numbers: Tel.:973-256-1699; Fax: 973-256-8341, E-mail: humanapr.com; or visit our Website: http//humanapress.com

Photocopy Authorization Policy:
Authorization to photocopy items for internal or personal use, or the internal or personal use of specific clients, is granted by Humana Press Inc., provided that the base fee of US $30 is paid directly to the Copyright Clearance Center at 222 Rosewood Drive, Danvers, MA 01923. For those organizations that have been granted a photocopy license from the CCC, a separate system of payment has been arranged and is acceptable to Humana Press Inc. The fee code for users of the Transactional Reporting Service is: [978-1-58829-858-4 $30].

e-ISBN: 978-1-59745-169-7

Printed in the United States of America. 10 9 8 7 6 5 4 3 2 1

Library of Congress Control Number: 2007938391

Preface

Over the past 20 years or so, there have been tremendous advances in our understanding of how normal cells transform to cancer and the importance of signaling pathways in cancer initiation and progression. This progress in scientific knowledge has resulted in the development of many drugs that target specific pathways with much wider therapeutic windows. In addition, technologic advances and the use of high-throughput approaches, particularly in the last 5 years, are beginning to impact the diagnosis and treatment of cancer. Microarray chips, for example, allow for global analysis of gene signatures in tumor and normal tissues; while proteomic approaches (functional—expression) allow for the evaluation of protein abundance and posttranslational modifications that are important in cancer development and progression. Thus genomics and proteomics approaches provide major opportunities in tumor diagnosis, in rational drug discovery and individualization of therapy for cancer patients. In spite of these advances in our knowledge and technical capabilities, there has not been a substantial decrease in overall death rate due to cancer. This is because too many cancers are diagnosed in late stage in the course of the disease. Therefore, we need an approach that will allow for early detection and proper diagnosis of the disease so it can be treated properly. For example, there is a need for proper use of human serum and other body fluids such as urine and sputum in early detection of cancer. Analyses of body fluids proteins by advanced proteomic profiling strategies allow for great opportunity for improved diagnostic and screening tests for early cancer. Furthermore, most pharmacological targets are proteins or DNA, not RNA. Any pharmacologist, given a chance, would choose information on protein expression over that RNA expression, so that is where we chose to start.

This book provides the reader with broad perspectives and breadth of knowledge on current topics related to the use of proteomic strategies in cancer therapy as well as anticipated challenges that may arise from its application in daily practice. The book is divided into four parts. The first part begins with the current technologies used in proteomics that allow for protein profiling and for the identification of druggable targets in human samples. Mass spectrometry-based protein characterization and protein microarrays hold great promise of predicting response to specific drugs in cancer therapy. The second part deals with the use proteomics in cell signaling. At present, the pharmaceutical and biotechnology industries have many potentially useful small molecule inhibitors of many pathways important in cancer that have yet to be taken to clinical trials. Understanding protein–protein interaction and posttranslational modifications through proteomics will likely make it much more feasible to do effective clinical trials of these small molecules alone and in combinations to overcome drug resistance and improve patient care. The third part of the book moves from signaling to actual clinical applications of proteomics in cancer therapy. Case studies in many tumor types are provided to show the feasibility of generating the critical information needed for individualization of therapy in cancer patients. The final part of the book provides in-depth information on annotating the human proteome and the role of Food and Drug Administration (FDA) in regulating the use of proteomics in cancer therapy. To functionally annotate the human proteome, the Swiss Institute for Bioinformatics (SIB) and the European Bioinformatics Institute (EBI) initiated a major effort to distribute to the scientific community highly integrated information on human protein sequences. This initiative, which is called the Human Proteomics Initiative (HPI), aims to provide for each known human protein a wealth of information including the description of its function, domain and protein family classification, subcellular location, posttranslational modifications, variants and similarities to other proteins. Integration of bioinformatics into clinical application of proteomics in cancer therapy is outlined as well as regulations and policy of commercial application of proteomics in patient care.

The chapters in this book collectively provide the current status of proteomics in cancer therapy. While the editor regularly expresses the sentiment that "proteomics holds great promise in individualization of medicine in cancer patients", the examples provided herein are testament to the fact that this promise has many challenges. Proteomics technologies must be improved for more global analysis of protein content of cells, tissues and body fluids, as well as the posttranslational modifications. Clinical proteomics studies require a large team effort. This team includes basic scientists, translational scientists, clinicians and computer scientists, and this has many challenges in our academic world. Finally, with the large volume of data generated, advances in informatics and data mining will be necessary to realize the full potential of proteomics in cancer detection, prognosis, diagnosis and therapy.

Preface

I wish to thank all the contributors to this volume for their effort, and I am grateful to them for taking the time and having the patience to disseminate the detailed information required in order that others can succeed in the application of this technology in their practice. I would also like to thank Beverly A. Teicher, the series editor, for her support and guidance, and Gina Impallomeni for editorial assistance and Rhukea Hussain in organization and development of this volume.

Lastly, special thanks to the anchors of my life, my wife Mariam and our children Eihab, Hany, and Sarah for their encouragement and pleasurable distraction. In the memory of my parents who believed in the power of learning and gave too much, I dedicate this book.

Sayed S. Daoud

Contents

Preface v
Contributors xi

PART I: PROTEOMICS TECHNOLOGIES
1 Current and Emerging Mass Spectrometry
  Instrumentation and Methods
  for Proteomic Analyses 3
  *Belinda Willard, Suma Kaveti,
  and Michael T. Kinter*

PART II: CELL SIGNALING PROTEOMICS
2 Integration of Genomics and Proteomics in Dissecting
  p53 Signaling 39
  *Kyunghee Lee, Tao Wang, Abdur Rehman,
  Yuhua Wang, and Sayed S. Daoud*

3 Proteomic Profiling of Tyrosine Kinases
  as Pharmacological Endpoints for Targeted
  Cancer Therapy 59
  Moulay A. Alaoui-Jamali and Devanand Pinto

PART III: TUMOR PROTEOMICS
4 Oncoproteomics for Personalized Management
  of Cancer 81
  K. K. Jain

5 Application of Serum and Tissue Proteomics
  to Understand and Detect Solid Tumors 101
  *Christina M. Annunziata, Dana M. Roque,
  Nilofer Azad, and Elise C. Kohn*

6 Insight on Renal Cell Carcinoma Proteome 121
  *Cecilia Sarto, Vanessa Proserpio, Fulvio Magni,
  and Paolo Mocarelli*

7 Proteomics in Lung Cancer 139
  M. A. Reymond, M. Beshay, and H. Lippert

8 Proteomic Strategies of Therapeutic Individualization and Target Discovery in Acute Myeloid Leukemia 161
Bjørn Tore Gjertsen and Gry Sjøholt

9 New Tumor Biomarkers: Practical Considerations Prior to Clinical Application...................... 189
Nils Brünner, Mads Holten-Andersen, Fred Sweep, John Foekens, Manfred Schmitt, Michael J. Duffy, on behalf of the EORTC PathoBiology Group

PART IV: BIOINFORMATICS AND REGULATORY ASPECTS OF PROTEOMICS

10 Annotating the Human Proteome: From Establishing a Parts List to a Tool for Target Identification 211
Rolf Apweiler and Michael Mueller

11 Regulatory Issues in the Co-Development of Oncology Drugs and Proteomic Tests: An Overview 237
Dave Li, Joseph Hackett, Maria Chan, Gene Pennello, and Steve Gutman

Index 259

Contributors

Moulay Alaoui-Jamali • *McGill University, Montreal, Canada*
Christina M. Annunziata • *National Cancer Institute, Bethesda, MD*
Rolf Apweiler • *European Bioinformatics Institute-EMBL, Cambridge, England*
Nilofer Azad • *National Cancer Institute, Bethesda, MD*
M. Beshay • *University of Magdeburg, Germany*
Nils Brünner • *Copenhagen University Hospital, Copenhagen, Denmark*
Maria Chan • *Food and Drug Administration, Rockville, MD*
Sayed S. Daoud • *Washington State University, Pullman, WA*
Michael J. Duffy • *Nuclear Medicine Department, St. Vincent's University Hospital, National University of Ireland, Ireland*
John Foekens • *Erasmus MedicalCenter Rotterdam, Rotterdam, The Netherlands*
Bjørn Tore Gjertsen • *University of Bergen, Bergen, Norway*
Steve Gutman • *Food and Drug Administration, Rockville, MD*
Joseph Hackett • *Food and Drug Administration, Rockville, MD*
Mads Holten-Andersen • *Copenhagen University Hospital, Copenhagen, Denmark*
K. K. Jain • *PharmaBiotech, Basel, Switzerland*
Suma Kaveti • *Lerner Research Institute, Cleveland, OH*
Michael T. Kinter • *Lerner Research Institute, Cleveland, OH*
Elise C. Kohn • *National Cancer Institute, Bethesda, MD*
Kyunghee Lee • *Washington State University, Pullman, WA*
Dave Li • *Food and Drug Administration, Rockville, MD*
H. Lippert • *University of Magdeburg, Germany*
Fulvio Magni • *Desio Hospital-University of Milano Bicocca, Milano, Italy*
Paolo Mocarelli • *Desio Hospital-University of Milano Bicocca, Milano, Italy*
Michael Mueller • *European Bioinformatics Institute-EMBL, Cambridge, England*
Gene Pennello • *Food and Drug Administration, Rockville, MD*
Devanand Pinto • *Dalhousie University, Halifax, Canada*

VANESSA PROSERPIO • *Desio Hospital-University of Milano Bicocca, Milano, Italy*
ABDUR REHMAN • *Washington State University, Pullman, WA*
M. A. REYMOND • *University of Magdeburg, Germany*
DANA M. ROQUE • *National Cancer Institute, Bethesda, MD*
CECILIA SARTO • *Desio Hospital-University of Milano Bicocca, Milano, Italy*
MANFRED SCHMITT • *Department of Obstetrics and Gynecology, Technical University of Munich, Munich, Germany*
GRY SJØHOLT • *University of Bergen, Bergen, Norway*
FRED SWEEP • *University Medical Center Nijmegen, Nijmegen, The Netherlands*
TAO WANG • *Washington State University, Pullman, WA*
YUHUA WANG • *Washington State University, Pullman, WA*
BELINDA WILLARD • *Lerner Research Institute, Cleveland, OH*

I | Proteomics Technologies

1 Current and Emerging Mass Spectrometry Instrumentation and Methods for Proteomic Analyses

Belinda Willard, Suma Kaveti, and Michael T. Kinter

CONTENTS

1. INTRODUCTION AND SCOPE
2. NEW MASS SPECTROMETRY INSTRUMENTATION
3. NEW ION FRAGMENTATION METHODS
4. NEW PROTEIN ANALYSIS METHODS THAT USE MASS SPECTROMETRY

SUMMARY

The dynamic nature of the mass spectrometry (MS) experiment creates many opportunities for innovation and advance. The new ion generation methods that merited the Nobel Prize were especially revolutionary examples, but mass analyzer design, ion chemistry, ion detection, and data analysis have all seen significant performance advances. In the area of proteomics, the key moment for any of these innovations and advances comes when they are transferred from MS laboratory to the biomedical research laboratory. Indeed, the past 5 or so years have seen key aspects of electrospray and

From: *Cancer Drug Discovery and Development
Cancer Proteomics: From Bench to Bedside*
Edited by: S. S. Daoud © Humana Press Inc., Totowa, NJ

matrix-assisted laser desorption/ionization mature to the point that MS is now ubiqitous part of life science research and experiments that were once cutting edge have become routine. Similarly, the new methods described in this chapter are also beginning to pass this transfer test to produce unique results in a variety of biological systems.

Key Words: Mass spectrometry; Ion Fragmentation; Proteomics; mass analysis.

1. INTRODUCTION AND SCOPE

Over the course of the 1990s, the use of mass spectrometry (MS) to sequence and identify proteins exploded (Figure line) in terms of both the utility of the techniques and the frequency of their application has created a new field of biological research termed "proteomics." In 2002, the scientific community's recognition of the importance of proteomics culminated in the awarding of the 2002 Nobel Prize in Chemistry to John Fenn of Virginia Commonwealth University in Richmond, Virginia, and Koichi Tanaka of Shimadzu Corporation in Kyoto, Japan, "for their development of soft desorption ionisation methods for MS analyses of biological macromolecules" *(1)*.

The primary proteomics experiment of 2002 followed a relatively straight-forward in-gel digestion to MS analysis to database search protocol that remains very effective today. A personal observation would be that the majority of mass spectrometer systems dedicated to this identification experiment in 2002 continue to carry it out effectively today and at the same or greater rate (samples per day). In fact, one can extend this observation and predict that the majority of instruments carrying out the gel band identification experiment today will still be running the experiment effectively in 2010. This prediction is based on the quality of the mass spectrometer systems that the manufacturers provide, the skills of the individuals operating the systems, and the unique utility of the band identification experiment in biomedical research.

The power of the band identification experiment, however, has provided a strong base for the continued development of protein analysis by MS. In many ways, the continued evolution of MS has become linked to advances in proteomics—advances in MS create new methods for proteomic experiments while new challenges in proteomics drive the development of new instruments and new techniques in MS.

The purpose of this chapter is to review the advances in MS, for application in proteomics, which have occurred since the 2002

Nobel Prizes were awarded, and go beyond this standard gel band identification experiment. The focus will be on three main areas of advancement: new MS systems, new ion fragmentation methods, and new experimental methods for protein analysis that use this growing set of MS tools. Our approach to this chapter is focused on introducing these selected topics to the reader, with key references to the literature, and putting them in the context of the experiments described in the rest of the chapters. This chapter is not intended to be a comprehensive review of all papers published in this time frame.

2. NEW MASS SPECTROMETRY INSTRUMENTATION

During the growth phase of proteomics, the primary MS systems used were electrospray ionization-ion trap or electrospray ionization-quadrupole-time-of-flight (Q-TOF) instruments for the protein sequencing experiments that used tandem MS and matrix-assisted laser desorption/ionization-time-of-flight (MALDI-TOF) systems for peptide mass mapping experiments. As the different instrument systems move forward, key parameters in the operation of a MS system continue to be sensitivity, resolution, mass accuracy, mass range, and scan speed. Sensitivity is an integral part of the power of MS. In the proteomics experiment, sensitivity translates into a combination of reduced sample requirements for the routine protein analysis experiments and/or the ability to see lower abundance proteins when given the same amount of sample. Resolution and mass accuracy improvements translate into more accurate (or more confident) identifications with fewer peptides. This ability can be particularly important as various experimental conditions, such as the analysis of complex mixtures, make the identification of proteins with only one peptide desirable and needed. Expanded mass ranges, in turn, have opened the sequencing experiment to larger peptides, including the technique of top-down sequencing where proteins are introduced into the mass spectrometer intact (without digestion) and fragmented in the gas phase for sequencing. Finally, increases in scan speed of the newer instruments, linked partly to their sensitivity, allow more high-quality spectra to be acquired in a single experiment which gives corresponding increases in the amount of information that is acquired.

The new instrumental approaches discussed below cover recent advances in mass spectrometer systems that have been incorporated into proteomic experiments. The specific systems covered are linear ion traps (LIT), Fourier transform (FT) instruments including both ion cyclotron resonance (ICR) and Orbitrap designs, and MALDI-tandem MS systems.

2.1. Linear-Ion Trap

Since their introduction in the late 1990s, ion trap mass spectrometers have arguably been the most widely used instrument system in proteomics *(2,3)*. The attractiveness of the ion trap is due to several factors: the sensitivity in the collision-induced dissociation (CID) mode, the ability to perform MS^n experiments, the ease of use, and the low cost of the instruments compared to other high-performance tandem MS^n systems. Other instrument features like data-dependent acquisition, in which ions for CID analysis are automatically selected in real-time by the instrument, also added to the power of these original ion trap systems.

The first generation ion trap systems use a trapping region located between a ring electrode and two capping electrodes. The shape of this region is evolved from quadrupole mass filters such that the instruments are now referred to as quadrupole ion traps (QIT) or three-dimensional ion traps, to distinguish them from the second generation LIT. The key difference between linear and QIT is the larger size of the trapping region and the larger number of ions that can be contained in a linear trap. This advantage is related to the need for ion traps to control the total number of ions entering the trapping region in order to limit the undesirable consequences of space charge effects that occur when too many ions are stored in a small area. These effects can lead to a distortion of the ion cloud and its movement within the mass spectrometer that can ultimately result in a loss of resolution, shifts in the measured m/z to give poor mass accuracy, and problems with sensitivity and dynamic range. A process known as automatic gain control measures and limits the number of ions put in a trap to avoid space charge effects, but has the deleterious practical effect of discarding ions and limiting the sensitivity of the experiment.

The performance advantages of the LIT are based on the increased storage capacity of the trap due to the increased volume, increased efficiency of ion transfer from the external ion source into the trap, and increased trapping efficiency *(4)*. These advantages result in an overall increase in sensitivity and dynamic range for the LIT compared to the conventional QIT system. For example, one group compared the detection limit between the two types of ion traps for a series of standard peptides and found the detection limit for the LIT to be approximately 600-fold better than the QIT (~500 zmol vs. ~300 amol) *(5)*. A second benefit involves the improvement of dynamic range. The MS dynamic range of the LIT was found to be approximately 40 times greater than that found for the QIT *(6)*, similar increases in dynamic range have also been reported for MS/MS spectra *(5)*.

Although it is difficult to argue against the importance of increased sensitivity in any MS experiment, it is possible that the most practically useful advantage of the LIT operation is the faster scan rates that it supports. The faster scan rate is a direct result of the ability to place more ions in the trap and the improved counting statistics those additional ions provides. The increased scan rate is especially useful for so-called shotgun or Mudpit proteomic experiments that directly characterize complex mixture of proteins. The shotgun sequencing experiment begins with the tryptic digestion of a complete protein mixture (i.e., a whole cell homogenate) with direct Liquid Chromatography (LC)-tandem MS analysis *(7)*. The Mudpit experiment adds multi-dimensional chromatography by incorporating some element of chromatographic fractionation of this complex protein digest before the LC-tandem MS experiment *(8)*. Each experiment uses data-dependent acquisition, and the success of these approaches relies heavily on the ability of the mass spectrometer to isolate and fragment as many peptide ions as possible over the course of the LC separation. One systematic comparison of the QIT and LIT instruments is in analyzing a complex peptide mixture from Jurkat T leukemia cells *(5)*. The authors found that the LIT was able to identify fourfold to sixfold more peptides and proteins in the complex mixture under identical experimental conditions. In addition, these investigators found that 70% of the doubly and triply protonated peptides gave better quality fragmentation spectra in the LIT compared to QIT, as measure by the Sequest correlation scores. In another comparative study, results obtained with a LIT instrument were compared to those from a Q-TOF instrument in an analysis of the yeast proteome *(9)*. When the different instruments were operated in a way that acquired the same number of CID spectra, the LIT was found to produce 21% more matched peptide spectra than the Q-TOF instrument, although no increase in protein identification was observed. However, when the scan speed advantage of the LIT was utilized, the instrument recorded more CID spectra, matched more peptides, and identified more proteins than the Q-TOF instrument. Several additional shotgun proteomic studies have been successfully carried out using LIT technology *(10,11)*.

Another major area of interest in proteomic studies is the identification of post-translational modifications. Several studies involving the identification of protein modifications have been performed utilizing the improved sensitivity and scan rates of the LIT instrument *(6–14)*. One study took advantage of the fast scan rates of the LIT in order to determine both the in vitro and in vivo phosphorylation sites of NFATc2 *(12)*. The LIT was operated under conditions to take 100 MS/MS scans per minute. The fast scanning rate of the instrument allowed the investigators to predict possible phosphorylation sites and to use selected ion fragmentation approaches to probe the sample digest for these phosphopeptides. In these experiments, an MS scan was followed by 30 MS/MS scans resulting in the identification of six in vitro phosphorylation sites in a single experiment. The investigators were also able to utilize the high sensitivity of the experiment to identify five in vivo phosphorylation sites within the same protein. In addition to the identification of phosphorylation sites, LIT instruments have been successfully used to identify other post-translational modifications including S-nitrosylation sites *(13)*, methylation *(6)*, acetylation *(14)*, and nitration and chlorination *(15)*.

A final feature of the LIT is the simple existence of two ends to the ion-trapping region. While less obvious then the performance aspects noted above, this new feature of the ion trap geometry may ultimately be appreciated as the most significant. Ions must always be injected from the ion source located at one end of the LIT, but the other end has been incorporated into at least two novel uses. The first use is to allow alternative ion excitation methods, like electron transfer dissociation (ETD) *(16)*, that are described in greater detail in Section 3 of this chapter. The second use is in hybrid instruments that add unique scan functions and high-resolution capabilities to the strengths of the ion trap system. The Q-Trap instrument, for example, combines the capabilities of the triple quadrupole instrument with the LIT (the term Q-Trap is a trademarked name for a specific hybrid triple quadrupole-LIT instrument from MDS Sciex) *(17)*. These instruments have three quadrupole regions. The first two are regions containing conventional quadrupoles that can be operated as mass filters or as collision cells. The final quadrupole can be operated as either a conventional quadrupole mass filter or a LIT. The use of these instruments in proteomic studies have utilized unique scanning capabilities that allow the identification of specific fragmentation patterns that are observed for phosphopeptides and glycosylated peptides *(18,19)*. Other hybrid instruments are coupling the LIT to high-mass resolution instrument such as a Fourier transform-ion cyclotron resonance (FT-ICR) or Orbitrap mass spectrometers *(20,21)*. These hybrid instruments add a high-resolution mass analyzer to the system, which significantly enhances the mass accuracy of the experiment. The FT-ICR and Orbitrap instruments are discussed in greater detail in the following sections.

2.2. Fourier Transform-Ion Cyclotron Resonance

The FT-ICR instrument (also often referred to generically as FT-MS) is a traditional, although sophisticated and expensive, mass spectrometer system used for protein analysis. Indeed, some of the earlier applications of protein sequencing with tandem MS used FT-ICR experiments *(22)*. FT-ICR MS has always been an exciting topic in field of MS instrumentation because the sensitivity, resolution, and mass accuracy in these instruments is the highest available among all the various instrument configurations.

The principle of ICR has been known since the late 1940s and was refined significantly in 1960s *(23,24)*. In the 1970s, FT technology was combined with ICR to begin to utilize the complete potential of the ICR technique *(25)*. ICR works on the relatively simple principle that ions in a strong magnetic field are constrained to circular orbits perpendicular to the field. To be trapped in this field and retained in the instrument, an ion must move in a manner in which the outward centrifugal force balances the inward magnetic force. As the outward centrifugal force of an ion is determined by a combination of mass, charge, and frequency of the rotation, all trapped ions with a given m/z must have a characteristic cyclotron frequency. In a simplified view, this frequency is recorded as an image current and converted to the m/z value. Because the recording of the image current is a nondestructive detection method, signal-to-noise ratio can be enhanced by averaging many cycles before the ions are lost from the cell. Finally, the geometry of the instrument does not place any specific molecular weight limit on the analyte ions. The net effect of the ICR approach is a unique combination of wide molecular weight range, high-sensitivity detection, and high m/z resolution *(26)*.

FT-ICR systems have several functional problems that have traditionally limited their use to a modest number of more advanced MS laboratories. One set of problems has been due to the large size of the instruments and the complexity of operating the high-field magnet. These problems were "solved" partly by the availability of actively shielded magnets, but mostly by the growing value of MS in protein analyses and the undisputed power of FT-ICR in those experiments. In other words, the benefits of these high performance systems have simply begun to outweigh the problems enough to lead more laboratories to their use. More difficult functional problems, however, have included the inter-related problems of space-charge effects and slow scan speeds. Specifically, to achieve the high m/z resolution, few ions are introduced into the ICR cell in order to avoid any space-charge effects. This small number of ions, in turn, requires longer acquisition times to achieve appropriate signals giving slow scan speeds. As a result, these problems are particularly difficult for experiments using liquid chromatography as a sample inlet due to the need to operate on the chromatographic time scale and the more variable amount of sample that is introduced into the mass spectrometer.

One approach to minimize these issues has been the development of hybrid systems that trap the ions prior to entry into the ICR cell. This approach began with various configurations of quadrupoles and has now been extended by placing a LIT in front of the ICR cell. This hybrid LIT-FT-ICR systems adds several intriguing capabilities to the FT-MS analysis. First, the ion trap provides the needed means for ion accumulation and control outside of the ICR that enhances sensitivity and helps manage space charge problems. Using this technology, routine femtomolar sensitivities of standalone FT-ICR instrument are improved to zeptomolar sensitivities in some cases *(6)*. Second, the high resolution–high mass accuracy of components of the FT-MS can be combined with the rapid scan rates in CID experiments of the LIT in a single experiment to give scientists an overall system with unique analytical power. For example, complex scan functions have been used to take advantage of the strength of both mass spectrometers simultaneously during a single LC-MS experiment (Fig. 1). A typical complex scan function would begin with a basic data-dependent approach. A survey scan is carried out to determine the ions seen at that point in the chromatogram. Data in the survey scan are then used to direct subsequent functions in both the FT-ICR and LIT areas. For example, (Fig. 1A), the FT-ICR can be directed to record narrow m/z range, high-resolution scans that measure the peptide molecular weights with accuracy in the parts-per-million (ppm) range while the LIT area is used to record the corresponding CID spectra. Other variations on this simultaneous use of both instrument components have also been reported *(20,27)*. Ultimately, each CID spectrum is combined with a high accuracy molecular weight measurement that is used to constrain

Chapter 1 / Current and Emerging Mass Spectrometry Instrumentation

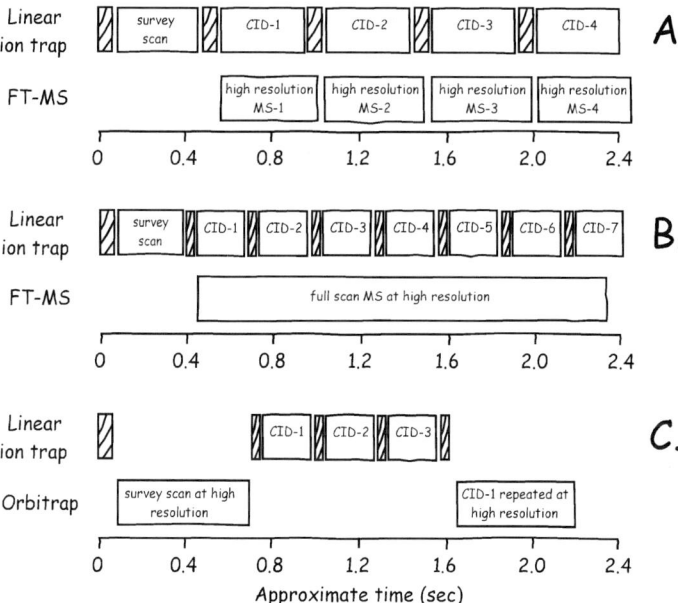

Fig. 1. Examples of possible scan functions that utilize both components of hybrid ion trap-high resolution mass spectrometry (MS) systems in a single analysis. The overall goal of these experiments is to acquire as many collision-induced dissociation spectra and the corresponding accurate peptides mass as possible on the chromatographic time scale. Panels A and B are based on descriptions given for combined linear ion trap-Fourier transform-MS (LIT-FT-MS) instruments (adapted from refs 20,27). Panel C is based on a description given for a combined LIT-Orbitrap instrument (adapted from ref. 21). The cross-hatched block represents time used for ion accumulation and automatic gain control. CID, collision-induced dissociation.

the database searches and increase the confidence of the protein identification (27).

The power of the accurate peptide molecular weight determination can be seen in the development of accurate mass tag (AMT) methods to quickly characterize proteomes (28,29). In this method, conventional HPLC-tandem MS methods are used to collect the sequence information of peptides and record their corresponding LC retention times. These peptides are considered as possible mass and time (PMT) tags. Their theoretical masses are calculated and confirmed on high-resolution FT-ICR by retaining the same sample preparation and liquid chromatographic conditions. Peptides that matched with the theoretical molecular masses are categorized as AMTs and are added to the PMT database. Subsequent analysis of such samples on the LC-FT-ICR system with searches of the PMT database, with stringent parameters, enables quick and confident identification of the source proteins.

2.3. Orbitrap

The Orbitrap mass spectrometer is based on the orbital trapping mass analysis concepts introduced by Kingdon in 1923 (The Orbitrap mass spectrometer is a trademarked name for a specific hybrid mass spectrometer from Thermo Fisher Corporation) *(30)*. This mass analysis technique utilizes two concentric electrodes, an inner electrode and outer electrode, on which voltages are applied to create a complex electrostatic field. When ions are injected into this field, they have two components to their movements—a rotation around the inner electrode that traps them in the device and an oscillation along the length of the trap at a frequency that is m/z dependent. As is the case in FT-ICR, the frequency of this oscillation is recorded as an image current and transformed from the time domain to the frequency domain to produce the mass spectrum. Key characteristics of these spectra are high mass resolution, in the range of 150,000, with correspondingly high mass accuracy, in the low ppm range *(31,32)*. One positive attribute of the Orbitrap, relative to FT-ICR, is a greater trap capacity before space charge effects become apparent. This capacity allows faster scan speeds in the high-resolution mode.

Like other trapping mass analyzers, ions are injected into the Orbitrap in pulses, trapped within the mass spectrometer, and detected. Unlike other trapping instruments, ion injection must occur in a very short time scale (<1 ms) due to high ion velocities and the absence of collisional cooling within the Orbitrap. In order to couple the Orbitrap mass spectrometer to a continuous ion source like electrospray ionization, additional ion storage and isolation steps must be used prior to this ion injection process. The first designs for the ion injection system used an rf-only quadrupole to accomplish this task with variable results. The most recent version uses a combination of a LIT and a C-trap prior to the Orbitrap *(33)*. The C-trap device is a type of rf-only quadrupole with curved rods that enhances the ability to accumulate and stores ions. The addition of the C-trap, combined with the use of a lock mass for mass calibration, has produced mass accuracies that are consistently better than 2 ppm *(34)*. As with the FT-ICR instruments noted above, this utilization of a LIT as part of the ion injection optics for the Orbitrap mass analyzer has also produced a hybrid mass spectrometer system in which complex scan functions can take advantage of ion dissociation in the LIT, and high resolution, high mass accuracy m/z measurements are made in the Orbitrap (Fig. 1).

The performance of the hybrid LIT-Orbitrap mass spectrometer in proteomic experiments was evaluated in the analysis of digested saliva samples *(21)*. For these experiments, the complex peptide mixture produced by the digestion of unfractionated salivary protein mixtures was analyzed by a shotgun approach that involved the direct LC-MS/MS analysis with the LIT-Orbitrap instrument. The LIT and Orbitrap mass analyzers can be used either simultaneously or independently. The experimental method involved a high-resolution mass scan on the Orbitrap mass analyzer portion of the instrument followed by three tandem mass spectrometry scans in the LIT mass analyzer and one spectrum where the MS/MS fragment ions are analyzed at high resolution in the Orbitrap analyzer. The authors reported a mean mass accuracy of 1.5 ppm over the course of the experiment. The authors were also able to compare the fragmentation spectra obtained in the LIT and Orbitrap analyzers and found that ion abundances were similar and that there appears to be no loss of fragment ions in the transfer process between the LIT and Orbitrap. The high mass accuracy measurements made on the LIT-Orbitrap instrument will lessen the likelihood of false positives in protein and peptide identification experiments.

2.4. MALDI-Tandem MS

One constant issue in proteomics, indeed in all analytical methods, is the speed of the experiment. In proteomics, applications like the comprehensive mapping of 2D gels can produce several hundred samples from a single experiment in a short period of time. In other situations, small numbers of samples from many different experiments can also produce high sample loads for analysis. One approach to increasing sample throughput, the traditional idea of automation, is a key part of improving the capacity of mass spectrometric analyses. The automation of liquid chromatography, for example, has a long history and is continually improving. In many ways, the relatively advanced state of LC automation has kept the LC-ESI-tandem MS experiment in the lead in proteomics field. It should be noted, however, that this type of automation does not actually increase the speed of the analysis but rather helps operate the instrument for longer periods of time.

MALDI analysis, on the other hand, is a fundamentally rapid experiment because the LC separation is either eliminated or greatly simplified (i.e., simple sample preparation methods like ZipTip cleanup). A basic mass mapping experiment with MALDI can be completed in a minute or two, as opposed to the hour needed for a standard LC-MS experiment. The problem with MALDI, however, has been the limitation of the mass mapping experiment, particularly as the interest in mixture analysis has grown. In fact, the appreciation that even simple samples tend to contain multiple proteins has hurt the utility of the mass mapping experiment. Simply put, scientists are either unhappy with the lack of fundamental amino acid data or require that data for other reasons. As a result, new MALDI instruments are using tandem MS approaches to acquire CID spectra for the peptide sequence analysis. These instruments range from the simple use of MALDI as an ion source for ion trap or Q-TOF instruments to the design of specific mass analyzer systems, like the TOF-TOF instrument, for use with MALDI sources.

The first versions of MALDI-Tandem MS experiments were published in the mid-1990s soon after the power of MALDI was appreciated *(35–37)*. These first applications used ion trap and FT-ICR mass spectrometers. One advantage cited in these first papers was the speed of the analysis, although the 5–10 samples per day is not rapid by current standards it was significantly faster than the one or two samples per day capacity of the LC-tandem MS experiments of the time *(38)*. These authors also recognized another advantage of the MALDI approaches, the very low sample consumption per laser shot *(35)* and the extended sample interrogation times this feature allows.

A significant problem for the first generation MALDI-tandem MS instruments was the relatively poor fragmentation characteristics of the singly charged ions produced by MALDI. This limited amount of sequence information is fundamentally related to the way peptides fragment under low-energy CID conditions and can be appreciated if one considers Wysocki's mobile proton theory of peptide fragmentation *(39,40)*. In this model, the single proton attached to MALDI-generated ions would have limited mobility because of its expected localization on strongly basic amino acids like arginine and a correspondingly limited ability to promote fragmentation. One approach to enhancing the sequence information that can be obtained from MALDI-generated peptide ions was to design the TOF-TOF instrument and use high energy collision conditions to induce the dissociation reactions *(41)*. The high-energy CID appears to promote more extensive fragmentation of the singly charged ions, producing more extensive peptide sequence information. In fact, de novo sequencing of MALDI-TOF-TOF spectra has been reported *(42)*. This more informative type of MALDI-Tandem MS experiment has been incorporated into effective protein identification routines *(43)*.

3. NEW ION FRAGMENTATION METHODS

A critical part of the protein identification experiment is the ability to fragment peptide ions in a way that reveals the amino acid sequence of that peptide. As with nearly all tandem MS experiments, this fragmentation has generally been accomplished by CID. Advantages of CID certainly include the ease of incorporating the technique into the tandem MS instruments for high-sensitivity experiments, but the highly informative pattern of fragmentation seen with peptide ions should also be appreciated. In fact, the uniquely informative fragmentation seen in the CID spectra of doubly charged peptide ions that are common in tryptic digests was recognized in the earliest examples of protein sequencing with electrospray ionization *(44)*.

Recently, two applications have re-energized the interest in improved methods to activate peptide ions and induce fragmentation—the characterization of post-translation modifications (particularly phosphorylation) and the direct sequencing of larger peptides including intact proteins (so-called top-down sequencing). Phosphorylated peptides represent a type of peptide ion that does not fragment in an optimum manner in a CID experiment. Other problematic peptides include glycosylated peptides and some peptides modified by small molecules like drugs or xenobiotic species.

The collisional activation of CID utilizes energy transfer pathways that operate through a slow distribution of the kinetic energy of the collision into vibrational energy in the ion. The rate (so-called slow heating) and extent of this distribution gives an element of preference for the more easily cleaved bonds. In most peptides, this preference is seen in wide variation in the abundance of the product ions in the CID spectrum, including high abundance fragment ions and fragment ions that are not observed. In general, this range of abundance is not a problem because sufficient sequence information is still present in the spectrum, combined with other peptides in the analysis, so that the database search programs can still effectively identify the proteins.

In the case of modified peptides, however, the amino acid side chain becomes a highly preferred fragmentation site such that the peptide ions tend to lose specific neutral species in a reaction that consumes all of the internal energy imparted by the collision. In these analyses, the CID spectra show a limited number of high abundance ions from the facile loss of that species, but no or limited ions that reveal the amino acid sequence. In the case of phosphopeptides, this behavior leads the a prominent loss of H_3PO_4 and a fragmentation pattern in which it is clear that a peptide is phosphorylated. Unfortunately, the underlying amino acid sequence information can be so limited that it is not possible to assign the peptide to a protein and/or determine the exact site of the modification. When attempting to fragment larger ions, including the direct analysis of intact proteins, the problem is slightly different and relates to basic limitations in adding sufficient energy to the ion to give effective fragmentation. Specifically, the collisional activation process for large ions creates a scenario where a large object is colliding with a small object. In this situation, a center-of-mass problem is created that limits the transfer of collision energy into the large ion. The second contributor to the problem is the larger number of bonds that must be broken. Ultimately, CID of large ions is plagued by a limited energy transfer going into too large a number of degrees of freedom and limited fragmentation is observed.

These problems have lead to two alternative types of ion activation—electron capture dissociation (ECD) *(45–47)* and ETD *(16,48)* that are rapidly being incorporated into commercially available systems. The key characteristic of these dissociation methods is the more direct activation of the amide bonds that produce a rapid fragmentation reaction at that site.

3.1. Electron Capture Dissociation

ECD was developed and described by Zubarev, McLafferty and co-workers in a pair of papers that appeared in the late 1990s *(45,46)*. The current excitement about this new ion dissociation technique relates the superior performance of ECD relative to CID, in these two problem experiments, the characterization of post-translational modifications and the direct sequencing of intact proteins.

The method of ECD uses the reaction of thermal electrons with positively charged species that are trapped in the ICR cell of an FT-MS system. For this experiment, electrons are produced by a filament or cathode, thermalized, and directed into the ICR cell. In the case of proteins and peptides, the recombination of the electron and the positive charge is modestly exothermic and produces an odd-electron ion that rapidly fragments through a reaction that breaks the N-αC bond, that is bond between the amide nitrogen and the alpha carbon of the peptide backbone (Fig. 2). This fragmentation reaction is unique, producing c- and z-ions, rather than the b- and y-ions seen in CID

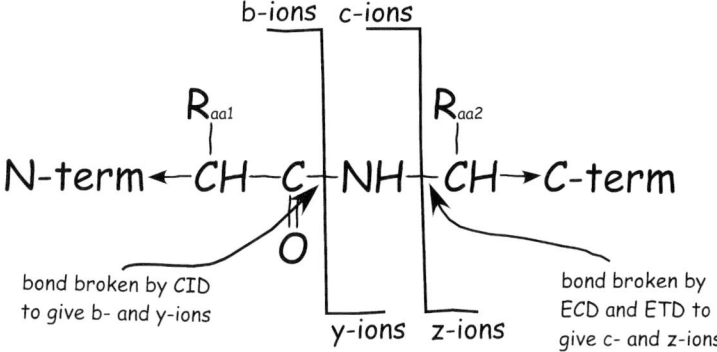

Fig. 2. A summary of the fragmentation pathways for electron capture dissociation (ECD) and electron transfer dissociation (ETD) vs. collision-induced dissociation (CID).

spectra based on breaking the carbonyl C–N bond. A key part of the electron capture reaction is that the fragmentation reaction occurs rapidly without redistribution of the energy. This trait produces a pattern of fragmentation that occurs only at the site of the recombination reaction. McLafferty refers to this as a non-ergotic process to reflect the fact that equilibration of the energy does not occur *(45)*. One result of this non-ergotic behavior is that ECD tends to not break side chain bonds, preserving post-translational modifications and leading to the enhanced utility for these types of studies. The second result of the non-ergotic behavior is that the direction of the fragmentation reaction is based on the site of the charges, such that the extent of fragmentation is not limited by the size of the ion. In fact, the multiple charging seen in electrospray ionization of proteins actually facilitates a more complete fragmentation by ECD.

The application of ECD is in relatively early stages of development, especially compared to the 40+ year history of CID. In addition, the development has been limited to some degree by the lack of commercial devices until approximately 2004. The applications that have been reported, however, clearly illustrate the potential of the technique. Initial published examples in the area of post-translational modifications have focused on the localization of phosphorylation sites *(49,50)*. These reports utilized the unique properties of ECD to generate product ion spectra without little or no side chain fragmentation, including no observable loss of H_3PO_4 from either the molecular ion or any of the fragment ions (Fig. 3). In addition, the utility for other

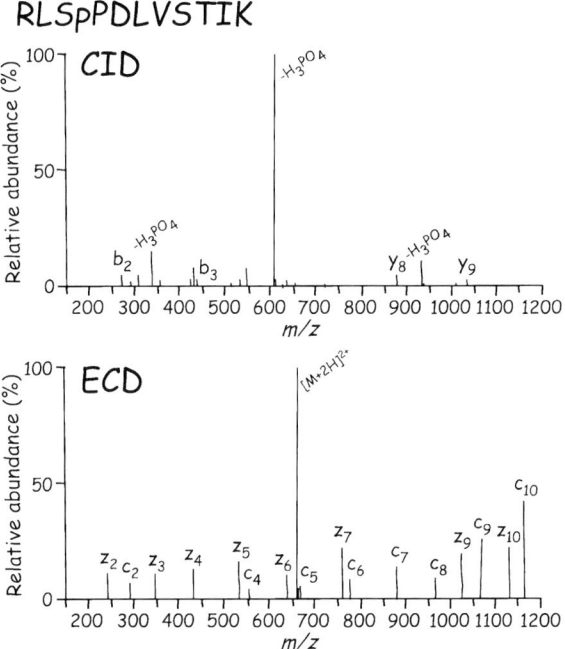

Fig. 3. A comparison of electron capture dissociation (ECD) and collision-induced dissociation (CID) spectra for a representative phosphopeptide. Representative spectra for the fragmentation of the hypothetical phosphopeptide RLSpPDLVSTIK (Sp denotes phosphoseriine). The CID spectrum is dominated by a doubly charged ion from the loss of H_3PO_4, seen as a loss of 49 Da from the doubly charged molecular ion. Limited additional fragmentation in the form of b- and y-ions is observed. The clarity of these data is diminished further by the confounding loss of H_3PO_4 from these fragment ions as well. The ECD spectrum, on the other hand, gives a clear set of c- and z- ions that reveal nearly the entire peptide sequence. No loss of H_3PO_4 is seen. The most abundant ion in the ECD spectrum is the unfragmented molecular ion.

types of modifications, including glycosylation *(51,52)*, carboxylation *(53)*, deamidation *(54)*, and sumylation *(55)*, have also been successfully determined. The other primary application, top-down sequencing, is described in greater detail below.

As the power of ECD is recognized, new fundamental studies are enhancing our understanding of the process *(56–58)*. Results of these fundamental studies will surely be an enhancement of the process, as exemplified by the design of multistep activation schemes to enhance the amount of sequence information *(59)*, and the improvement of ECD-based instruments *(60,61)*, as exemplified by the reports of the use of ECD on ion trap instruments *(62,63)*. Finally, it is notable that the more complete fragmentation characteristics of ECD are also being applied to the "standard" peptide sequencing experiment *(64)*. Recently, advances in the speed of the ECD process have allowed its incorporation into data-dependent experiments used for the routine protein identification experiment *(65)*.

3.2. Electron Transfer Dissociation

ETD was introduced by Hunt and Co-workers in 2004 *(16,48)*. The electron transfer method is closely related to the electron capture method. Indeed, all of the downstream actions of the process relating to how the protein and peptide ions fragment are seemingly identical to those described above for electron capture.

The key difference between electron transfer and electron capture is, as implied by the names, how the electron is placed on the ion that is being dissociated. In the electron transfer process, this process begins with a reagent ion that is reminiscent of chemical ionization reagent ions. In fact, in the first description of ETD, the reagent ion, a mixture of anthracene anions, was produced in a chemical ionization source with methane and anthracene *(16)*. These reagent ions are then injected into an ion trap mass spectrometer to react with the m/z-selected ion of interest and the product ion spectrum recorded in a standard tandem MS experiment.

As is the case with ECD, the ion/ion reaction of ETD is exothermic and the excess energy is used by the peptide or protein ion to break the backbone N–αC bond between the amide nitrogen and the alpha carbon of the peptide backbone to produce are c- and z-ions (Fig. 2). Again, no distribution of the energy takes place, so the site of the fragmentation is determined by site of the charge. Because these sites tend to be randomly distributed in electron ionization, a population of ions tend to have all sites charged to at least some extent, and an extensive series of the c- and z-type product ions is produced.

A key distinction between ETD and ECD is the types of instruments on which the experiments can be performed. The electron capture process is not suitable for instruments that use rf fields, like ion trap instruments, because of problems with the inability to trap thermal electrons in these fields and the overall residency time of the ions in the trap. As a result, ECD is best suited to ICR instruments. In contrast, the negative ions used in ETD can be trapped in LIT instruments by taking advantage of the two entrance points and the multiple segments in these mass analysers *(16)*.

4. NEW PROTEIN ANALYSIS METHODS THAT USE MASS SPECTROMETRY

With the combination of new mass analyzers, new ion excitation methods, and the continued strengths of the traditional systems, investigators have been able to devote considerable energy extending the uses of MS in proteomics to go well beyond the standard protein identification experiment. Three directions merit particular attention: the continued development of the direct analysis approaches (shotgun sequencing) that broadly identify specific subclasses of proteins, the development of quantitative methods, and the development of methods that work directly with intact proteins in the mass spectrometer. The shotgun sequencing experiments are a group of direct analysis techniques that are used in situations where the power of the mass spectrometer adds sensitivity, dynamic range, and/or speed to an experiment to create a more comprehensive experiment. These types of experiments began with the Mudpit experiments noted above *(8)*, but also include experiments that retain some sample clean-up or separation by gel electrophoresis separation (so-called GeLC experiments) *(66,67)*. A notable new application of the shotgun experiment is in the area of phosphoproteomics. In quantitative analyses, new methods are being reported that utilize the inherent relationship between the strength of a MS signal and the amount of analyte present in the sample. Many of the approaches use different types of labeling techniques [i.e., metabolic labeling, isotope-coded affinity tag (ICAT), and iTRAQ], but label-free methods are also being reported. Finally, new MS experiments are using ECD and ETD described above to directly sequence proteins. These top-down sequencing experiments are a very promising development, partly because of the simple power of a more direct analysis (i.e., no gel and no digestion) and partly because they can provide new information about aspects of the protein structure that are not predicted based on the gene sequence (i.e., post-translational modification, post-translational processing, and splice variants). As a group, these new techniques promise to add significant new capabilities to the next generation of proteomic experiments.

4.1. Phosphoproteomics

Protein phosphorylation is one of most physiologically important post-translational modifications due to its role in the signaling cascades that regulate many biological processes. First generation experiments were focused on individual target proteins and sought to characterize the sites of phosphorylation so that subsequent experiments could methodically address the significance of each site using techniques like site-directed mutagenesis. For this approach, the protein of interest is typically expressed in a tagged form (e.g., FLAG- or myc-tagged), pulled from the expression system with immobilized antibodies to the tag, digested with a protease, and analyzed by data-dependent LC-tandem MS methods *(68)*. The expression system is generally treated with phosphatase inhibitors and may be either stimulated or have the candidate kinase co-expressed to enhance the amount of phosphorylation. A few important factors make the characterization of phosphorylation site significantly more difficult than the identification experiment: the low frequency of the modification (a stoichiometry problem), limitations in the amount of sequence information obtained from the target protein (a coverage problem), and issues with the sequence information given in a CID spectrum (the neutral loss problem). Nonetheless, this standard approach has been very productive and will continue to add to our basic understanding of protein phosphorylation. At the same time, this target-by-target, site-by-site approach has distinct limitations. Not only is the pace of progress relatively slow, but the isolated approach does not recognize the breadth of these pathways, their inter-connected nature, and the need to include quantitative information. As a result, new phosphoproteomic experiments are beginning to use shotgun approaches, with quantitation, to address these issues.

The most difficult part of characterizing protein phosphorylation is the initial step of finding the relevant ions in complex data sets. These datasets contain what literally amounts to tens of thousands of the detected ions. Furthermore, the abundance-based logic used in data-dependent analyses is distinctly biased against the low abundance signals of which the phosphopeptides are a component. Therefore, the most rational approach to solving this problem is to enrich the digest in phosphopeptides. The two main approaches being used are immobilized metal affinity chromatography (IMAC) and strong cation exchange (SCX).

The oldest approach, IMAC, uses the affinity of the phosphate group for certain metal ions as a way to pull the phosphopeptides out of a mixture *(69)*. Traditionally, Fe^{3+} was used for this purpose but other reports with Ga^{3+}, Ni^{2+}, and Cu^{2+} have also been made. In some instances, enhanced selectivity of certain metal ions has been demonstrated. For example, Ga^{3+} IMAC was used in combination with phosphatase treatments to detect several phosphorylation sites in c-Jun NH_2^- terminal kinase (JNK)-interacting protein 1 *(70)*. Overall, however, the general success of IMAC technology has been inconsistent and of relatively limited value for most users trying to study isolated proteins.

The IMAC experiment, however, has appeared significantly more robust when applied to mixtures in broader phosphoproteome experiments. The first report came from the laboratory of Hunt and co-workers *(71)*. This experiment used an Fe^{3+} IMAC column to isolated phosphopeptides from a yeast protein preparation. A key alteration of the standard IMAC protocol was the inclusion of a methylation step that esterified the acidic amino acids and reduced non-specific binding of peptides containing aspartate and glutamate residues *(71)*. This modification also gave an opportunity to add a stable-isotope label (D_3-methyl) for quantitative analyses. This initial report of phosphoproteome analysis described the detection of 216 phosphopeptides and with specific assignment of 383 phosphorylation sites in the yeast proteome *(71)*. In a subsequent set of experiments, this same technique was used in a human system, sperm cells *(72)*. This IMAC-based phosphoproteome approach has also been refined to specifically target tyrosine phosphorylation *(73,74)*. These experiments use anti-phosphotyrosine antibodies to immunoprecipate this sub-class of proteins. The precipitate is digested, the peptides methylated, and the phosphorylated peptides isolated from the mixture with either Ga^{3+} or Fe^{3+} IMAC columns. An initial report identified over 70 tyrosine-phosphorylated peptides in a single analysis *(74)*.

The second approach, SCX, takes advantage of the fundamental acidity of phosphopeptides to separate these peptides from the larger pool of unmodified peptides. The first report by Gygi and co-workers used an SCX system to characterize the phosphoproteome of human cells in culture (Hela cells) *(75)*. In this analysis, the proteins were digested with trypsin, the digest applied to the SCX column, and the phosphopeptides eluted in the early part of an NaCl gradient. The elution order was multiply phophorylated peptides eluting first, the singly phosphorylated peptides eluting second, and the unphosphorylated peptides eluting last. A total of 967 phosphoproteins were identified through the characterization of 2002 phosphorylation sites *(75)*. Similar experiments were used to extend the method to animal tissues with equally impressive results *(76)*.

The success of these phosphoproteome experiments, whether based on IMAC or SCX, has produced a quantum leap forward in the number phosphorylation sites that are now known. As a group, it is interesting to consider the amount of information that is present in this series of just three papers in human systems—the identification of several thousand phosphorylation sites (ignoring for the sake of argument the likely overlap between parts of the datasets) *(72,74,75)*. The most direct utilization of these data is the larger scale evaluation of phosphorylation motifs *(77)*. It should be remembered, however, that these data also represent unique resources for the functional studies of each protein in the dataset. Not only do they provide new information about the sites of the phosphorylation, but they would also give useful practical information about the LC-MS characteristics of the specific peptides. Finally, as discussed below, these phosphoproteome methods are also being combined with quantitative methods to more completely address the dynamics of these systems.

4.2. Quantitative Techniques

A key piece in the continued development of new proteomic methods has been the introduction of viable approaches to making quantitative determinations. As a group, these methods build on the traditional role of MS for quantitative analyses in studies of small molecules like drugs and metabolites. Although the intensity of mass spectrometer signal is proportional to the amount of analyte present in a sample, the exact details of this proportionality are always complex and can be affected by many experimental factors, including the chemical characteristics of the analyte, variations in the sample preparation procedure, and daily (or even hourly) differences in instrument performance. As a result, quantitative methods that use MS generally incorporate some kind of internal standard against which the analyte signal is normalized. For small molecule analyses, these internal standards tend to be analogs of the analyte that have either a different molecular weight or different chromatographic retention time that allows the internal standard signal to be recorded independently of the analyte signal.

In proteomic applications, an issue for internal standard design is the protein digestion step and the dramatic changes the characteristics of the analyte (i.e., from a protein to a series of peptides) that it produces. The effect of this change is to create a challenging question of how best to incorporate an internal standard into the analysis? A second, unique aspect of quantitative analyses in proteomic experiments is that the absolute amount of a given protein analyte is usually far less important then any change in the amounts between two or more states. As a result, the quantitative methods used in proteomics have largely focused on comparative analyses for the specific task of finding proteins with either increasing or decreasing expression between different experimental conditions, but little interest in the amount of unchanged proteins. The approaches to these comparative analyses can be broadly classified as either labeling methods, that add a distinguishing label to the proteins and peptides from each source, or label-free methods, that take this relative quantity information directly from the MS dataset. A variety of methods exist and these methods are described in more detail below. It is important to note that all of these methods have been used by different investigators to study complex biological systems with good results. It is fair to observe, however, that while this range of quantitative methods gives one a useful set of options from which to choose, no method has distinguished itself as far superior to the others. An interesting conceptual question, as one considers use of the various labeling versus label-free methods, is "what is the role of the proteomic experiment in the overall investigation of a biological system?" A personal opinion would be that if the results are intended to be evaluated and presented as a "stand-alone result", then the labeling methods may be preferable because they give clear, numerical results that lend themselves to the types of statistical evaluation that generally accompanies quantitative analysis. If, on the other hand, the results are used in a discovery context in which additional levels of experimentation will be done, such as Western blot analysis or ELISA, then the label-free methods may be preferable because they are easier and less expensive to carry out, yet still provide effective leads to direct those subsequent experiments.

4.2.1. LABELING METHODS

The incorporation of differentiating labels into proteins and peptides is accomplished in two general ways—either by metabolic labeling of the protein prior to isolation and digestion or by chemical derivatization of the peptides after protein isolation and digestion (Fig. 4).

The first metabolic labeling method was an adaptation of the ^{15}N-labeling method used by structural biologists to label proteins for nuclear magnetic resonance (NMR) *(78,79)*. Separate cultures of yeast were grown in media that contained either ^{14}N or ^{15}N in the

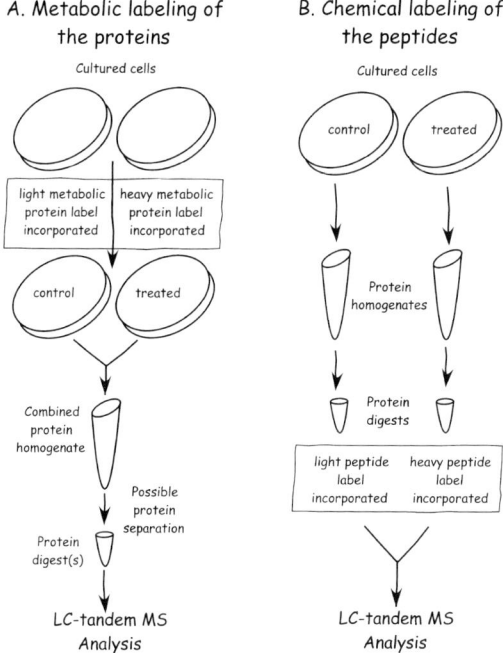

Fig. 4. A summary of labeling methods used for quantitative proteomic experiments. Examples of two general approaches are shown. (A) For the metabolic labeling of the proteins, cultured cells are grown in the presence of a suitable labeling reagent (i.e., ^{15}N-ammonium or D_6-arginine) which is incorporated into cellular protein as it is synthesized during cell growth over several passages. The labeled cells are then treated and processed to isolate the protein, and aliquots of protein from each source are mixed. The mixed protein samples are the digested for analysis. It may be advantageous to separate the protein mixture by electrophoresis prior to the digestion. (B) For the chemical labeling of peptides, cells are grown and treated under standard culture conditions. The proteins from each source are isolated and digested separately, before labeling with the respective labeling reagents. After labeling, aliquots of the digest from each source are mixed and analyzed.

available nitrogen sources, such that the respective isotopic forms of the nitrogen were uniformly incorporated to label all the proteins in that culture as either light (^{14}N-containing) or heavy (^{15}N-containing), respectively. After exposing one of the cultures to the treatment being studied, each culture (control and treated) was harvested, mixed in equal proportions, and the mixed set of proteins run in a 2D gel *(78)* or by shotgun methods *(79)*. For each protein that was identified, the detected peptide or peptides also reflected the abundance of the parent protein in the ratio of the heavy form of the peptide to the light form. These experiments are somewhat restricted by the ability to limit the nitrogen source to suitable pure ^{14}N or pure ^{15}N reagents, for example, isotopically pure ammonium sulfate *(79)*. As a result, the published applications have tended to be in yeast, although whole animal labeling of rats has also been reported *(80)*.

A more recent approach with wider applicability, including mammalian cell culture systems, is known as stable isotope labeling by amino acids in culture (SILAC) *(81)*. The SILAC experiments uses isotopically labeled essential amino acids as the labeling source. The original report, for example, used leucine in either the unlabeled (D_0) or deuterium-labeled (D_3) form *(81)*, but more recent applications appear to favor various labeled lysine or arginines *(82,83)*. In all cases, the basic rationale of the approach and experimental design is identical to that described above for the ^{14}N/^{15}N labeling. Specifically, different cultures of cells are maintained in either the light or heavy isotope, treated, harvested, and mixed to give a sample in which the heavy-to-light ratio reflects the relative amount of the protein. An interesting expansion of the metabolic labeling strategy that the SILAC method allows is the ability to make multiple comparisons through the use of multiple labels. For example, a SILAC approach with unlabeled arginine (Arg0), arginine labeled to be +6 Da (Arg6), and arginine labeled to be +10Da (Arg10) was used to compare protein expression in stem cells treated with either platelet-derived growth factor (PDGF) or epidermal growth factor (EGF) *(82)*. The three labels allowed multiple comparisons, including EGF treatment to PDGF treatment and each treatment to the untreated control. Overall, a survey of the literature reveals a far more extensive use of the SILAC method, compared to the ^{14}N/^{15}N metabolic labeling method.

The first of the peptide derivatization methods was the ICAT method of Aebersold and co-workers *(84)*. The original ICAT reagent was a pair of either D_0- or D_8-labeled cysteine-reactive reagents that labeled all cysteine-containing peptides in a digest. The key feature of the ICAT reagent was that it also included a biotin group, so that the labeled peptides incorporated both isotope-labeling and an affinity for avidin-type columns. As the method was intended for use in shotgun identification experiments in complex mixtures, this affinity functionality gave an effective means for simplifying the mixture and, ideally, increasing the number of different proteins that could be sampled. The biotin-containing peptides (also containing the isotope-coding) are bound to a strepavidin column to separate them from the large pool of unlabeled peptides. Again, each pair of peptides detected carries two pieces of information about the parent protein—the identity, based on the CID spectrum, and the relative amounts in the two systems, based on the D_8/D_0 ratio. A subsequent version of the ICAT reagent added ^{13}C-labeling and a cleavable biotin moiety *(85)*. These changes were incorporated to enhance the performance of the ICAT method by eliminating the chromatographic separation of the heavy- and light-labeled peptides that can be observed with highly deuterated species, and improving the identification process by eliminating the undesirable effects on the CID spectrum of the large biotin-containing side group. The elegance and power of the ICAT method has inspired a number of alternative approaches to incorporating isotope codes into peptides. These methods include other cysteine-specific labels, N-terminal labels, and C-terminal labels *(86–88)*.

One of the newer methods to peptide labeling that has attracted considerable attention is the amine-reactive isobaric-tagging reagents of Pappin and co-workers (marketed as iTRAQ reagents) *(89)*. The molecular design of these reagents has three parts, an succinimide head group for facile addition to amines, an isotope-coded reporter group at the tail that provides the relative quantity information, and a central mass balance region that compensates for the variable mass of the report and equalize to overall mass of a labeled peptide. Indeed, the key to the reagent is this combined effect of the isotope-coded reporter region and the mass balance region. When added to a peptide, these reagent parts give a common addition of 145 Da, regardless of the respective isotope-coded reporter region. When fragmented, however, the isotope coding is seen in the low m/z region of the CID spectrum. The elegance of the method is the way that the different reporter groups are incorporated without adding to the complexity of the initial peptide complex, because each peptide remains at a single molecular weight. A total of four reagents are currently available, with reporter groups for m/z 114, 115, 116, and 117, to accommodate experimental designs needing multiple comparisons.

4.2.2. LABEL-FREE METHODS

Investigators are also realizing that sufficient quantitative information can be obtained directly from the LC-MS experiment, without the additional effort or expense of the peptide labeling steps. These so-called label-free methods use a bioinformatic approach to compare elements of raw LC-MS datasets, such as signal strength or the number of characterized peptides from a given protein, to make informative comparisons between samples.

As noted above, a key barrier to quantitation with MS involves the fact that while the intensity of an ion does reflect its abundance, corrections must be made for differences in mass spectrometer response, sample loading, and ionization efficient that can occur between experiments. The foundation of most label-free methods is the recognition that most complex samples have a large group of unchanging species that can be used as the source of normalizing factors (Fig. 5). These normalizing factors are used to correct the sample-to-sample and

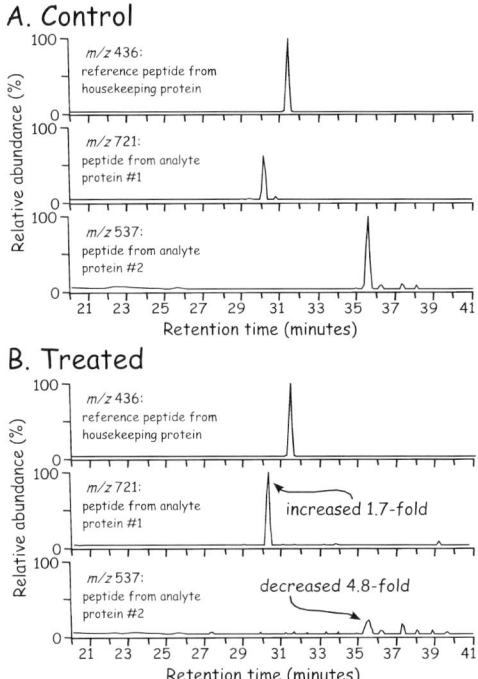

Fig. 5. The general approach to label-free quantitative analyses. The experiment begins with the selection of appropriate peptides representative of each protein of interest in the analysis. This selection process includes the selection of a suitable reference protein with unchanging expression to be used as a "loading control." All signals are normalized to this control and changes in expression determined based on the normalized response for each protein monitored. It may be advantageous to use several peptides per protein if possible. These data may be taken from raw LC-mass spectrometry datasets or selectively acquired using techniques like selected reaction monitoring.

run-to-run variation in signal intensities for the species of interest. This normalization approach has been validated in at least two studies; by spiking in known amounts of myoglobin in human plasma and by adding various biomarkers to human serum *(90,91)*. Both of these studies showed a linear response in ion intensity with increasing concentration of the analytes in a reproducible fashion.

Two approaches to label-free quantitation have been used; a signal intensity approach and a spectrum counting approach. The most common method for determining signal intensity for a given analyte is to integrate the chromatographic peak area in an LC-MS experiment *(92)*. One challenge to this process is the variability in retention time that is seen in most low flow systems used for proteomic analysis. Several reports have used computer programs to analyze the LC-MS chromatograms to match up the elution profiles prior to peak area determination *(93,94)*. Once the peak areas are determined, a reference is chosen and each peak area is expressed as a ratio to the reference sample. It has been found that the results are independent of the choice of the reference sample. A recent report studied several different complex proteomes including rat kidneys, human plasma, and colon tumor cell lines. Multiple LC-MS runs were performed and the results indicate that peptide elution profiles and signal intensities were highly reproducible, and this is reflected in 90% of the peptide ion ratios deviated less than 20% from the average *(93)*.

The second type of quantitation involves relating protein abundance directly to the number of peptide CID spectra obtained in the LC-MS/MS experiment. This approach is based on work in yeast that showed that the probability of a protein being identified is related to its abundance *(95)*. The more abundant proteins resulted in more fragmentation spectra than lower abundance proteins. The authors used the number of spectra acquired for each protein, termed spectral count, as a measure of its relative abundance. A correlation between spectral counts and abundance was observed with high linearity *(95)*. A comparative study of peak intensity measurements and spectral count measurements for protein quantitation has been performed *(96)*. Both the peak intensity and spectral count measurements resulted in both linearity and reproducibility using standard proteins, and both methods were validated with independent measurements using gel staining followed by sodium dodecyl sulfate–polyacrylamide gel electrophoresis. Indeed, these authors noted that the two techniques were complementary with spectral counting method being more sensitive because the same peptides did not need to be observed in each LC-MS dataset, but the peak ion intensity methods being more accurate because signal intensity is more directly related to amount of analyte *(96)*.

As with the peptide-labeling methods above, label-free methods have also been effectively applied to quantitative analysis of post-translational modifications. In its simplest form, the quantitation of post-translational modifications can be performed by normalizing the peak areas of modified peptides to the peak areas of another, unmodified peptide from the protein being studied *(97,98)*. The usefulness of this technique is dependence of the proper selection of a reference peptide that is not being modified. To lessen this issue, methods that use multiple peptides for the normalization have also been reported *(99)*.

4.3. Top-Down Sequencing

As noted above, new ion activation methods like ECD and ETD give an effective means of fragmenting larger ions, including intact proteins. This new capability has lead to the new term of "top-down sequencing" that distinguishes these experiments from the peptide-based sequencing or "bottom-up" experiments. One possible advantage of a top-down approach would be that the more direct analyses are generally easier and faster to carry out, because of fewer manipulations and no digestion. However, this advantage is unlikely to apply in the case of routine protein identification by top-down analysis because the bottom-up approach is simply too effective. Instead, the real advantage of top-down sequencing is the important types of new information that preserved and accessible in an intact protein. In particular, any variation between the database sequence that is based on DNA sequencing and the translated form that exists in vivo. This information may indeed exist in the digested protein sample, but the complexity of these datasets and the reliance of the data analysis tools (the search programs) on comparing the data to the DNA sequence-derived database sequence means this information is either missed or ignored. As a result, top-down sequencing is particularly valuable in studies of post-translational processing of proteins, including post-translational modification and proteolytic processing.

The top-down experiment works because the fragmentation process in ECD and ETD is directed by the sites of charge recombination reactions between the protonated molecular ions and thermal electrons. As noted above, these reactions give rapid, local energy transfer that drive an amide bond fragmentation reaction that is less limited by the size of the ion being fragmented and is more evenly distributed throughout the structure of the ion. The products of the fragmentation reactions are still highly charged, so the high resolution of the FT-ICR instruments that supports the electron capture process is also needed to reveal the charge state of each fragement ion and give interpretable ECD spectra.

In the original top-down experiment, the $[M+11H]^{11+}$ ion of ubiqutin was fragmented by ECD to cleave 50 of 75 amide bonds *(45)*. Other examples in this initial report included 40 of 103 amide bonds in the $[M+15H]^{15+}$ ion of cytochrome C and 33 of >250 amide bonds in the $[M+21H]^{21+}$ ion of carbonic anhydrase. Interestingly, this carbonic anhydrase example also illustrated one on the limits in top-down sequencing, because ECD of the $[M+34H]^{34+}$ ion gave only lower charge state molecular ions but no fragment ions *(45)*. One explanation for the more limited fragmentation of the larger proteins was the effect of intramolecular non-covalent interactions that increase in complexity as protein size increases. These effects were lessened by adding collisional activation to the ECD process and produces more than threefold increases in the amounts of sequence information obtained in the analysis of proteins up to approximately 40 kDa *(59)*. This series of papers ultimately lead to the use of ECD to obtain other detailed structural information from an intact protein, including the absence of an N-terminal Met and disulfide bond formation by top-down analysis of a series of proteins with molecular weights up to 42 kDa *(100)*. Since that time, the use of top-down analysis to do more than identify proteins has gone is many directions, including the characterization of post-translational modification and splice variation *(101–104)*. These types of experiments will likely be the most significant contribution of the top-down approach to proteomic analyses.

REFERENCES

1. Nobelprize.org. Nobel prize in Chemistry 2002.
2. Stafford, G. Jr. J. Am. Soc. Mass Spectrom. 13:589–596, 2002.
3. Jonscher, K.R., Yates, J.R. III. Anal. Biochem. 244:1–15, 1997.
4. Schwartz, J.C., Senko, W.W., Syka, J.E. J. Am. Soc Mass Spectrom. 13:659–669, 2002.
5. Viveka, M., Rezaul, K., Cong, Y.S., Han, D. Mol. Cell. Proteomics 4:214–223, 2005.
6. Syka, J.E., Marto, J.A., Bai, D.L., Horning, S., Senko, M.W., Schwartz, J.C., Ueberheide, B., Garcia, B., Busby, S., Muratore, T., Shabanowitz, J., Hunt, D.F. J. Proteome Res. 3:621–626, 2004.
7. McCormack, A.L., Schieltz, D.M., Goode, B., Yang, S., Barnes, G., Drubin, D., Yates J.R. III. Anal. Chem. 69:767–776, 1997.
8. Washburn, M.P., Wolters, D., Yates, J.R. III. Nat. Biotechnol. 19:242–247, 2001.
9. Elias, J.E., Haas, W., Faherty, B.K., Gygi, S.P. Nat. Methods 2:667–675, 2005.

10. Plymoth, A., Yang, Z., Lofdahl, C.G., Ekberg-Jansson, A., Dahlback, M., Fehniger, T.E., Marko-Varga, G., Hancock, W.S. Clin. Chem. 52:671–679, 2006.
11. Garcia, B.A., Smalley, D.M., Cho, H., Shabanowitz, J., Ley, K., Hunt, D.F. J. Proteome Res. 4:1516–1521, 2005.
12. Villar, M., Ortega-Perez, I., Were, F., Cano, E., Redondo, J.M., Vazquez, J. Proteomics 6:S16–S27, 2006.
13. Garcia, B.A., Busby, S.A., Barber, C.M., Shabanowitz, J., Allis, C.D., Hunt, D.F. J. Proteome Res. 3:1219–1227, 2004.
14. Martinez-Ruiz, A., Villanueva, L., Gonzalez de Orduna, C., Lopez-Ferrer, D., Higueras, M.A., Tarin, C., Rodriguez-Crespo, I., Vazquez, J., Lamas, S. Proc. Natl. Acad. Sci. U. S. A. 102:8525–8530, 2005.
15. Zheng, L., Settle, M., Brubaker, G., Schmitt, D., Hazen, S.L., Smith, J.D., Kinter, M. J. Biol. Chem. 280:38–47, 2005.
16. Syka, J.E., Coon, J.J., Schroeder, M.J., Shabanowitz, J., Hunt, D.F. Proc. Natl. Acad. Sci. U. S. A. 101:9528–9533, 2004.
17. Le Blanc, J.C., Hager, J.W., Ilisiu, A.M., Hunter, C., Zhong, F., Chu, I. Proteomics 3:859–869, 2003.
18. Amoresano, A., Monti, G., Cirulli, C., Marino, G. Rapid Commun. Mass Spectrom. 20:1400–1404, 2006.
19. Sandra, K., Devreese, B., Van Beeaumen, J., Stals, I., Claeyssens, M. J. Am. Soc. Mass Spectrom. 15:413–423, 2004.
20. Olsen, J.V., Mann, M. Proc. Natl. Acad. Sci. U. S. A. 101:13417–13422, 2004.
21. Yates, J.R. III, Cociorva, D., Liao, L., Zabrouskov, V. Anal. Chem. 78:493–500, 2006.
22. Hunt, D.F., Shabanowitz, J., Yates, J.R. III, Zhu, N.Z., Russell, D.H., Castro, M.E. Proc. Natl. Acad. Sci. U. S. A. 84:620–623, 1987.
23. Hipple, J.A, Sommer, H., Thomas H.A. Phys. Rev. 76:1877–1878, 1949.
24. Anders, L.R., Beauchamp, J.L., Dunbar, R.C., Baldeschwieler, J.P. J. Chem. Phys. 45:1062–1065, 1966.
25. Comisarow, M.P., Marshall, A.G. Chem. Phys. Lett. 25:282–283, 1974.
26. Bogdanov, B., Smith, R.D. Mass Spectrom. Rev. 24:168–200, 2005.
27. Haas, W., Faherty, B.K., Gerber, S.A., Elias, J.E., Beausoleil, S.A., Bakalarski, C.E., Li, X., Villen, J., Gygi, S.P. Mol. Cell. Proteomics 5:1326–1337, 2006.
28. Strittmatter, E.F., Ferguson, P.L., Tang, K., Smith, R.D. J. Am. Soc Mass Spectrom. 14:980–991, 2003.
29. Adkins, J.N., Monroe, M.E., Auberry, K.J., Shen, Y., Jacobs, J.M., Camp, D.G. II, Vitzthum, F., Rodland, K.D., Zangar, R.C., Smith, R.D., Pounds, J.G. Proteomics 5:3454–3466, 2005.
30. Kingdon, K.H. Phys. Rev. 21:408–418, 1923.
31. Makarov, A. Anal. Chem. 72:1156–1162, 2000.
32. Hu, Q., Noll, R.J., Makarov, A., Hardman, M., Cooks, R.G. J. Mass Spectrom. 40:430–443, 2005.
33. Makarov, A., Denisov, E., Kholomeev, A., Balschun, W., Lange, O., Strupat, K., Horning, S. Anal. Chem. 78:2113–2120, 2006.
34. Olsen, J.V., de Godoy, L.M., Li, G., Macek, B., Mortensen, P., Pesch, R., Makarov, A., Lange, O., Horning, S., Mann, M. Mol. Cell Proteomics 4:2010–2021, 2005.
35. Qin, J., Ruud, J., Chait, B.T. Anal. Chem. 68:1784–1791, 1996.
36. Qin, J., Chait, B.T. Anal. Chem. 69:4002–4009, 1997.

37. Carroll, J.A., Penn, S.G., Fannin, S.T., Wu, J., Cancilla, M.T., Green, M.K., Lebrilla, C.B. Anal. Chem. 68:1798–1804, 1996.
38. Qin, J., Fenyo, D., Zhao, Y., Hall, W.W., Chao, D. M., Wilson, C.J., Young, R.A. Chait, B.T. Anal. Chem. 69:3995–4001, 1997.
39. Wysocki, V.H., Tsaprailis, G., Smith, L.L., Breci, L.A. J. Mass Spectrom. 35:1399–1406, 2000.
40. Tabb, D.L., Smith, L.L., Breci, L.A., Wysocki, V.H., Lin, D., Yates J.R., III. Anal. Chem. 75:1155–1163, 2003.
41. Medzihradszky, K.F., Campbell, J.M., Baldwin, M.A., Falick, A.M., Juhasz, P., Vestal, M.L., Burlingame, A.L. Anal. Chem. 72:552–558, 2000.
42. Yergey, A.L., Coorssen, J.R., Backlund, P.S. Jr., Blank, P.S., Humphrey, G.A., Zimmerberg, J., Campbell, J.M., Vestal, M.L. J. Am. Soc. Mass Spectrom. 13:784–791, 2002.
43. Bienvenut, W.V., Deon, C., Pasquarello, C., Campbell, J.M., Sanchez, J.C., Vestal, M.L., Hochstrasser, D.F. Proteomics 2:868–876, 2002.
44. Covey, T.R., Huang, E.C., Henion, J.D. Anal. Chem. 63:1193–1200, 1991.
45. Zubarev, R.A., Kelleher, N.L., McLafferty, F.W. J. Am. Chem. Soc. 120: 3265–3266, 1998.
46. Zubarev, R.A., Kruger, N.A., Fridricksson, E.K., Lewis, M.A., Horn, D.M., Carpenter, B.K., McLafferty, F.W. J. Am. Chem. Soc. 121:2857–2862, 1999.
47. Cooper, H.J., Hakansson, K., Marshall, A.G. Mass Spectrom. Rev. 24:201–222, 2005.
48. Coon, J.J., Ueberheide, B., Syka, J.E., Dryhurst, D.D., Ausio, J., Shabanowitz, J., Hunt, D.F. Proc. Natl. Acad. Sci. U .S. A. 102:9463–9468, 2005.
49. Stensballe, A., Jensen, O.N., Olsen, J.V., Haselmann, K.F., Zubarev, R.A. Rapid Commun. Mass Spectrom. 14:1793–1800, 2000.
50. Shi, S.D., Hemling, M.E., Carr, S.A., Horn, D.M., Lindh, I., McLafferty, F.W. Anal. Chem. 73:19–22, 2001.
51. Renfrow, M.B., Cooper, H.J., Tomana, M., Kulhavy, R., Hiki, Y., Toma, K., Emmett, M.R., Mestecky, J., Marshall, A.G., Novak, J. J. Biol. Chem. 280:19136–19145, 2005.
52. Hakansson, K., Cooper, H.J., Emmett, M.R., Costello, CE., Marshall, A.G., Nilsson C.L. Anal. Chem. 73:4530–4536, 2001.
53. Kelleher, N.L., Zubarev, R.A., Bush, K., Furie, B., Furie, B.C., McLafferty, F.W., Walsh, C.T. Anal. Chem. 71:4250–4253, 1999.
54. Cournoyer, J.J., Lin. C., O'Connor, P.B. Anal. Chem. 78:1264–1271, 2006.
55. Cooper, H.J., Tatham, M.H., Jaffray, E., Heath, J.K., Lam, T.T., Marshall, A.G., Hay, R.T. Anal. Chem. 77:6310–6319, 2005.
56. Syrstad, E.A., Turecek, F. J. Am. Soc Mass Spectrom. 16:208–224, 2005.
57. Breuker, K., Oh, H., Lin, C., Carpenter, B.K., McLafferty, F.W. Proc. Natl. Acad. Sci. U. S. A. 101:14011–14016, 2004.
58. McFarland, M.A., Chalmers, M.J., Quinn, J.P., Hendrickson, C.L., Marshall, A.G. J. Am. Soc. Mass Spectrom. 16:1060–1066, 2005.
59. Horn, D.M., Ge,Y., McLafferty, F.W. Anal. Chem. 72:4778–4784, 2000.
60. Patrie, S.M., Charlebois, J.P., Whipple, D., Kelleher, N.L., Hendrickson, C.L., Quinn, J.P., Marshall, A.G., Mukhopadhyay, B. J. Am. Soc. Mass Spectrom. 15:1099–1108, 2004.
61. Jebanathirajah, J.A., Pittman, J.L., Thomson, B.A., Budnik, B.A., Kaur, P., Rape, M., Kirschner, M., Costello, C.E., O'Connor, P.B. J. Am. Soc. Mass Spectrom. 16:1985–1999, 2005.

62. Baba, T., Hashimoto, Y., Hasegawa, H., Hirabayashi, A., Waki, I. Anal. Chem. 76:4263–4266, 2004.
63. Silivra, O.A., Kjeldsen, F., Ivonin, I.A., Zubarev. R.A. J. Am. Soc. Mass Spectrom. 16:22–27, 2005.
64. Axelsson, J., Palmblad, M., Hakansson, K., Hakansson, P. Rapid Commun. Mass Spectrom. 13:474–477, 1999.
65. Cooper, H.J., Akbarzadeh, S., Heath, J.K., Zeller, M.J. Proteome Res. 4:1538–1544, 2005.
66. Rappsilber, J., Ryder, U., Lamond, A.I., Mann, M. Genome Res. 12:1231–1245, 2002.
67. Schirle, M., Heurtier, M.A., Kuster, B. Mol. Cell. Proteomics 2:1297–1305, 2003.
68. Schroeder, M.J., Webb, D.J., Shabanowitz, J., Horwitz, A.F., Hunt, D.F. J. Proteome Res. 4:1832–1841, 2005.
69. Andersson, L., Porath, J. Anal. Biochem. 154:250–254, 1986.
70. D'Ambrosio, C., Arena, S., Fulcoli, G., Scheinfeld, M.H., Zhou, D., D'Adamio, L., Scaloni, A. Mol. Cell. Proteomics 5:97–113, 2006.
71. Ficarro, S.B., McCleland, M.L., Stukenberg, P.T., Burke, D.J., Ross, M.M., Shabanowitz, J., Hunt, D.F., White, F.M. Nat. Biotechnol. 20:301–305, 2002.
72. Ficarro, S., Chertihin, O., Westbrook, V.A., White, F., Jayes, F., Kalab, P., Marto, J.A., Shabanowitz, J., Herr, J.C., Hunt, D.F., Visconti, P.E. J. Biol. Chem. 278:11579–11589, 2003.
73. Salomon A.R., Ficarro, S.B., Brill, L.M., Brinker, A., Phung, Q.T., Ericson, C., Sauer, K., Brock, A., Horn, D.M., Schultz, P.G., Peters, E.C. Proc. Natl. Acad. Sci. U. S. A. 100:443–448, 2003.
74. Brill, L.M., Salomon, A.R., Ficarro, S.B., Mukherji, M., Stettler-Gill, M., Peters, E.C. Anal. Chem. 76:2763–2772, 2004.
75. Beausoleil, S.A., Jedrychowski, M., Schwartz, D., Elias, J.E., Villen, J., Li, J., Cohn, M.A., Cantley, L.C., Gygi, S.P. Proc. Natl. Acad. Sci. U. S. A. 101:12130–12135, 2004.
76. Ballif, B.A., Villen, J., Beausoleil, S.A., Schwartz, D., Gygi, S.P. Mol. Cell Proteomics 3:1093–1101, 2004.
77. Schwartz, D., Gygi, S.P. Nat. Biotechnol. 23:1391–1398, 2005.
78. Oda, Y., Huang, K., Cross, F.R., Cowburn, D., Chait, B.T. Proc. Natl. Acad. Sci. U. S. A. 96:6591–6596, 1999.
79. Washburn, M.P., Koller, A., Oshiro, G., Ulaszek, R.R., Plouffe, D., Deciu, C., Winzeler, E., Yates, J.R., III. Proc. Natl. Acad. Sci. U. S. A. 100:3107–3112, 2003.
80. Wu, C.C., MacCoss, M.J., Howell, K.E., Matthews, D.E., Yates J.R., III. Anal. Chem. 76:4951–4959, 2004.
81. Ong, S.E., Blagoev, B., Kratchmarova, I., Kristensen, D.B., Steen, H., Pandey, A., Mann, M. Mol. Cell Proteomics 1:376–386, 2002.
82. Kratchmarova, I., Blagoev, B., Haack-Sorensen, M., Kassem, M., Mann M. Science 308:1472–1477, 2005.
83. Ong, S.E., Kratchmarova, I., Mann, M. J. Proteome Res. 2:173–181, 2003.
84. Gygi, S.P., Rist, B., Gerber, S.A., Turecek, F., Gelb, M.H., Aebersold, R. Nat. Biotechnol. 17:994–999, 1999.
85. Li, J., Steen, H., Gygi, S.P. Mol. Cell Proteomics 2:1198–1204, 2003.
86. Olsen, J.V., Andersen, J.R., Nielsen, P.A., Nielsen, M.L., Figeys, D., Mann, M., Wisniewski, J.R. Mol. Cell Proteomics 3:82–92, 2004.

87. Zhang, X., Jin, Q.K., Carr, S.A., Annan, R.S. Rapid Commun. Mass Spectrom. 16:2325–2332, 2002.
88. Yao, X., Freas, A., Ramirez, J., Demirev, P.A., Fenselau, C. Anal. Chem. 73:2836–2842, 2001.
89. Ross, P.L., Huang, Y.N., Marchese, J.N., Williamson, B., Parker, K., Hattan, S., Khainovski, N., Pillai, S., Dey, S., Daniels, S., Purkayastha, S., Juhasz, P., Martin, S., Bartlet-Jones, M., He, F., Jacobson, A., Pappin, D.J. Mol. Cell Proteomics 3:1154–1169, 2004.
90. Bondarenko, P.V., Cheliu, D., Shaler, T.A. Anal. Chem. 74:4741–4749, 2002.
91. Wang, W., Zhou, H., Lin, H., Roy, S., Shaler, T.A., Hill, L.R., Norton, S., Kumar, P., Anderle, M., Becker, C.H. Anal. Chem. 75:4818–4826, 2003.
92. Ono, M., Shitashige, M., Honda, K., Isobe, T., Kuwabara, H., Matsuzuki, H., Hirohashi, S., Yamada, T. Mol. Cell Proteomics 5:1338–1347, 2006.
93. Wang, G, Wu, W.W.,, Zeng, W., Chou, C.L., Shen, R.F. J. Proteome Res. 5:1214–1223, 2006.
94. Beck, H.C., Nielsen, E.C., Matthiesen, R., Jensen, H.L., Sehested, M., Finn, P., Grauslund, M., Hansen, A.M., Jensen, O.N. Mol. Cell Proteomics 5:1314–1325, 2006.
95. Liu, H., Sadygov, R.G., Yates, J.R., III. Anal. Chem. 76:4193–4201, 2004.
96. Old, W.M., Meyer-Arendt, K., Aveline-Wolf, L., Pierce, K.G., Mendoza, A., Sevinsky, J.R., Resing, K.A., Ahn, N.G. Mol. Cell Proteomics 4:1487–1502, 2005.
97. Ruse, C.I., Willard, B.B., Jin, J.P., Haas, T., Kinter, M., Bond, M. Anal. Chem. 74:1658–1664, 2002.
98. Willard, B.B., Ruse, C.I., Keightley, J.A., Bond, M., Kinter, M. Anal. Chem. 75:2370–2376, 2003.
99. Steen H., Jebanathirajah J.A., Springer M., Kirschner M.W. Proc. Natl. Acad. Sci. U. S. A. 102:3948–3953, 2005.
100. Ge, Y., Lawhorn, B.G., ElNaggar, M., Strauss, E., Park, J.H., Begley, T.P., McLafferty, F.W. J. Am. Chem. Soc. 124:672–678, 2002.
101. Medzihradszky, K.F., Zhang, X., Chalkley, R.J., Guan, S., McFarland, M.A., Chalmers, M.J., Marshall, A.G., Diaz, R.L., Allis, C.D., Burlingame, A.L. Mol. Cell Proteomics 3:872–886, 2004.
102. Siuti, N., Roth, M.J., Mizzen, C.A., Kelleher, N.L., Pesavento, J.J. J. Proteome Res. 5:233–239, 2006.
103. Thomas, C.E., Kelleher, N.L., Mizzen, C.A. J. Proteome Res. 5:240–247, 2006.
104. Roth, M.J., Forbes, A.J., Boyne, M.T, II., Kim, Y.B., Robinson, D.E., Kelleher, N.L. Mol. Cell Proteomics 4:1002–1008, 2005.

II Cell Signaling Proteomics

2 Integration of Genomics and Proteomics in Dissecting p53 Signaling

Kyunghee Lee, Tao Wang, Abdur Rehman, Yuhua Wang, and Sayed S. Daoud

CONTENTS

1. INTRODUCTION
2. GENOMICS OF P53 SIGNALING
3. PROTEOMICS AND P53 TARGET IDENTIFICATION
4. INTEGRATION OF PROTEOMICS IN P53 SIGNALING
5. CONCLUSIONS

SUMMARY

The discovery of the human genome and subsequent expansion of proteomics research combined with emerging technologies such as sophisticated computational biology are producing unprecedented changes in our understanding of the role of tumor suppressors in cell signaling. p53, a key tumor suppressor gene, is mutated in the majority of human cancers. Successful outcome of chemotherapy and radiotherapy, in many cases, depends on functional p53. Therefore, elucidation of the function, regulation, and molecular interactions of p53 with its targets is of great importance for developing successful cancer therapy. Because p53 is a transcription factor with an expanded repertoire of genes that are known to be directly

From: *Cancer Drug Discovery and Development
Cancer Proteomics: From Bench to Bedside*
Edited by: S. S. Daoud © Humana Press Inc., Totowa, NJ

or indirectly under its control, global analysis of gene expression profiles represents the best approach for studying the p53 response to chemotherapeutic agents. Our laboratory and others have used gene expression profiling with microarrays to isolate p53 target genes in an attempt to understand the molecular signaling of p53 and to determine the molecular consequences that follow from treatment of cancer cells with chemotherapeutic agents. Although these attempts have been successful in isolating novel genes, gene expression profiling alone at the transcription level is not sufficient. Most of pharmacological targets are proteins or DNA, not RNA. Therefore, we need an approach that focuses on the molecular profiling of cancer cells at the DNA, RNA, and proteins. In this chapter, key findings on the current genomics and proteomics approaches for isolating p53 targets and their functional analysis are reviewed, and perspective is provided on the potential of integrating both approaches in the molecular pharmacology of p53 signaling.

Key Words: Pharmacogenomics; proteomics; p53; S100A4; RB18A; colon cancer

1. INTRODUCTION

The discovery of the human genome *(1,2)* and subsequent expansion of proteomics research with emerging technologies such as sophisticated computational biology are producing unprecedented changes in our understanding of systems biology. The measurement of genes and proteins has gained increasing acceptance as means by which to study the response of an organism to stressful conditions, whether they are environmental, genetic, pharmacological, or toxicological. In the context of pharmacology, these measurements referred are to as pharmacogenomics and proteomics, which undoubtedly have provided new biological insight that was not attainable a decade ago. They have also been impacting on our future direction in drug discovery and individualization of therapy. Nonetheless, integration of information obtained from genomics and proteomics is desirable as it links the individual biological elements together to provide a more complete understanding of dynamic biological processes. In addition to developing new data mining methods (bioinformatics) to extract further details from each of the multidimensional datasets produced by many genomic and proteomic approaches, effort is also being expanded to improve statistical analysis, as indicated in our recent genomics study *(3)*, searchable databases *(4,5)*, annotations, and molecular interaction mapping *(6)* to maximize our learning of cell signaling. Thus, integration of information obtained from genomics and proteomics study would also provide many significant types of scientific resources absent from the typical hypothesis-driven study such as new methodology, a database resource, and a hypothesis generator *(7)*.

There are several recent examples, in both mammalian and nonmammalian systems, in which genes and proteins have been integrated using either biology- or data-driven strategies *(8–11)*. However, this chapter addresses the evolving field of genomics, proteomics, and bioinformatics in dissecting p53 signaling for molecular targeting discovery. The p53 protein is a central transcription factor activated in response to a variety of cellular stresses, including DNA damage, mitotic spindle damage, heat shock, metabolic changes, hypoxia, viral infection, and oncogene activation *(12,13)*. p53 can induce growth arrest and apoptosis, events that prevent the survival of damaged cells. The transactivation function of p53 is mediated through sequence-specific binding of its central domain to cis-acting elements within the promoters or introns of responsive genes. Many genes are known be activated by p53, most of them in growth arrest or apoptotic pathways *(14)*. Consequently, the downstream effects of activating p53 are complex, and no single pathway mediates the full range of functions of p53. Indeed, recent observations regarding the existence of alternate splice variants of p53 *(15)*, the complexity of p53 regulation *(16)*, and the existence of allelic variants of p53 and its regulators with distinct functionality *(17)* make the situation even more complex. Thus, newer strategies including analysis of the expression of downstream targets of p53 using integrated approaches of genomics and proteomics may provide more robust measures of the p53 signaling pathway.

2. GENOMICS OF P53 SIGNALING

In the past decade, numerous efforts were made to identify p53 target genes through various gene expression techniques, including microarrays expression analysis and integrative genomic assays. However, these technologies have provided limited coverage for p53 bindings due to the unavailability of the full sequences of the human genome. Following the discovery of the human genome, many novel p53 target genes were identified due to the scientific advancement in genomic methodologies coupled with the availability of many sophisticated molecular database resources. To date, additional p53 responsive genes are identified mostly based on the chromatin immunoprecipitation (ChIP) DNA fragment pool, including

ChIP-quantitative polymerase chain reaction (ChIP-qPCR), ChIP-microarrays chip (ChIP-on-chip), ChIP paired-end ditag (ChIP-PET), serial analysis of binding elements (SABE), serial analysis of gene expression (SAGE), and RNA interference (RNAi) knockdown. The following is the review on the current status of these methodologies and their use in the identification of p53 target genes. Some perspectives for their use in analyzing the p53 signaling pathway are also discussed.

2.1. Microarray-Based Transcript Profiling

Microarray technology represents a rapid, semiquantitative system for gene expression profiling. In the context of p53 signaling, many laboratories have used genomic microarrays for the identification of p53 primary and secondary target genes under inducible conditions of p53 protein *(18–20)*. To analyze the p53 dependence of molecular events after stressful response such as DNA damage, we compared gene expression changes in a p53 wild-type human colon carcinoma cell line, HCT-116 (p53+/+), with those in an isogenic p53 knockout (p53–/–) after treatment with the topoisomerase I inhibitor topotecan *(3)*. To simultaneously study the effects of p53 status and topotecan treatment at different concentrations and time points, we developed a new experimental design for microarray studies. We term it a cross-referenced network design (Fig. 1). The most frequently used design for two-color microarray experiments simply compares each sample with a single internal reference sample by cohybridization. Our network design used internal reference sample, but it also provided the additional global set of comparison indicated in Fig. 1. Based on this design we were able to analyze the entire network of data by multiple regression in an approximately weighted fashion to increase the statistical reliability of the results and conclusions in terms of the statistical consistency test. This approach led to the selection of 167 genes based on both p53 and treatment dependence. The data were displayed using a novel two-dimensional visualization map called gene expression map (GEM). The GEM (Fig. 2) shows how the 167 differentially expressed genes can be grouped on the basis of treatment effect and p53 effect. By analogy with cartographic depictions of the world, the roles of longitude and latitude are played by p53 effect and treatment effect, respectively. Genes found in a particular region of the map have similar expression profiles and may also be associated in a more fundamental, physiological way, perhaps in some cases occurring in a common pathway. For example, four genes in the transforming growth factor-β

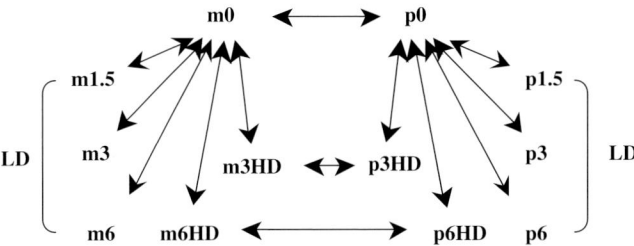

Fig. 1. Cross-referenced network. Total RNA from p53 wild-type (positive) cells [p, HCT-116 (p53+/+)] and p53 knockout (minus) cells [m, HCT-116 (p53–/–)] were collected at the beginning of treatment ($t = 0$) or at the indicated time point ($t = 1.5$, 3, or 6 h) after 1 h of treatment with either a low (LD = 0.1 µM) or higher concentration (HD = 1 µM) of topotecan. As indicated by arrows, pairs of conditions were directly compared by labeling test samples with Cy3 or Cy5 and reference samples with Cy5 or Cy3 and by cohybridization equal amounts of the product to a single array slide. The arrows are double-headed to indicate that hybridizations were done in pairs with reversal of the colors to eliminate any dye-related bias. Initially, all times and doses were compared with the $t = 0$ samples. Cell types (p, m) were then compared at $t = 0$, 3, and 6 h at HD. The network of comparisons was used mathematically (by weighted multiple linear least-squares regression) to calculate the relative expression differences between all pairs of conditions, not just those cohybridized on the same array. Expression was assessed relative to m0 for subsequent analysis. [Reproduced with permission from Daoud et al. 2006 *(3)*, copyright 2003, Cancer Research].

(TGF-β) signaling pathway appeared in the NE quadrant of the GEM (Fig. 2). The up-regulation of these genes was p53 dependent. This induction of the TGF-β pathway provides an example of the indirect effects of p53 on gene transcription by activation of other transcription factors. The assignment of S100A4 transcript in far Eastern region (E) of the GEM provides a testable hypothesis that the expression of S100A4 transcript is associated with the expression of p53 in cells. The p53 dependency for S100A4 gene expression in colorectal p53+/+ was also confirmed by other laboratory *(21)*. The role of S100A4 in p53 signaling is further discussed in section that outlines the use of integrated proteomics in p53 signaling.

2.2. Chromatin Immunoprecipitation-Based Assays

The complete sequence of the human genome has allowed us to identify many of the binding sites for a specific transcription factor. ChIP assay has been successfully used for this purpose. It is a technique that extracts DNA fragments to which a transcription factor is bound.

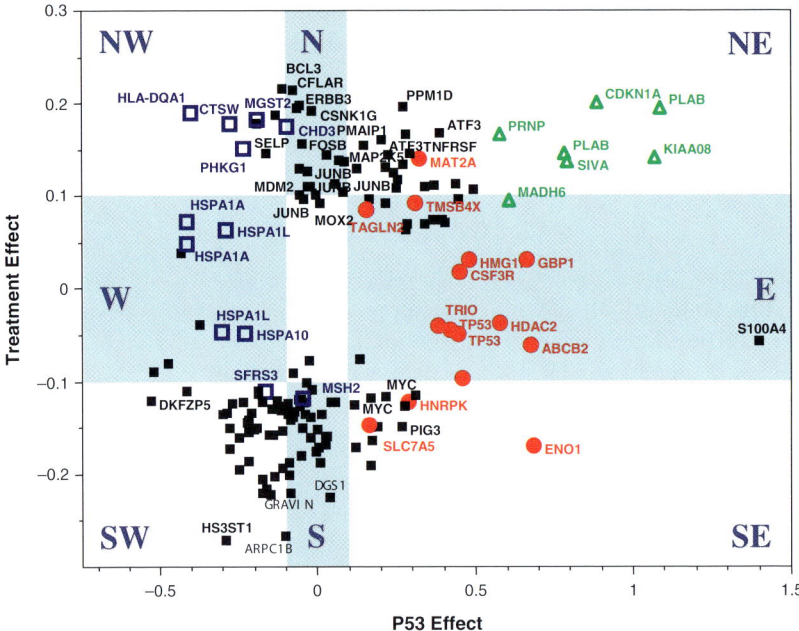

Fig. 2. Gene expression map (GEM). Genes were mapped according to the effects of p53 status (horizontal axis; longitude) and treatment (vertical axis; latitude) on their expression levels with reference to untreated p53 cells. The origin point (0,0) corresponds, therefore, to a gene whose expression is unaffected by p53 status or treatment. Each axis shows $\log_{10}$ of the contribution of each effect to the observed overall response (expression difference). For example, a gene located at $X = +0.3$, $Y = +0.3$ is expressed at twofold higher level in p53+/+ cells than in p53–/– cells and an additional twofold higher level in response to treatment with topotecan. This result is about a fourfold increase when treated p53+/+ cells are compared with untreated p53–/– cells. The GEM is divided into eight regions containing genes with roughly the same qualitative behavior in terms of p53 status and treatment response. For convenience, regions are labeled according to compass directions (N, NE, E, SE, S, SW, W, and NW). [Reproduced with permission from Daoud et al. 2006 *(3)*, copyright 2003, Cancer Research].

One approach to identify these fragments is to hybridize them to microarrays (ChIP-on-chip), which has been successfully used in yeast *(22)* but mammalian genome proved to be too large for this approach. Alternatively, the fragments can be sequenced, but this has proved cumbersome for mammalian genomes, and only a few chromosomes have been studied in any one screen *(23,24)*. To scan the entire genome for binding sites, a new sequencing approach that is based on ChIP assay called ChIP-PET was introduced *(25)*. To identify p53-binding sites, the authors used the PET sequencing approach. In this method, after cloning the precipitated fragments with ChIP, the 5´ and 3´ ends of several clones are concatenated for efficient sequencing. Each pair ends are then mapped to the genome to identify a potential binding site. As with our previous study *(3)*, the authors here used the wild-type HCT-116 human colon carcinoma cells and an isogenic p53 knockout counterpart after treatment with 5-fluorouracil to activate p53 expression. After a comparison with expression data, the authors of this study identified 98 novel genes that are direct targets of p53 and 24 previously identified genes. They also found a difference in the position of the binding site between genes that were up-regulated and down-regulated and even identified a second p53-binding site in the promoter of a well-known p53-target gene, *CDKN1A* (*aka* p21, Cip1). In fact, our genomics study *(3)* identified this gene, and its expression is influenced by both p53 and drug treatment as shown in the NE region of the GEM (Fig. 2). Several of the newly identified p53 targets such as *S100A2* are involved in cell motility, which is interesting as p53 is involved in suppressing tumor metastasis. Absent, however, is the identification of *S100A4* previously identified by our study? This may indicate that *S100A4* gene is not a direct target for p53 and that there is no p53-binding site in the promoter of *S100A4* gene. Therefore, additional studies with ChIP-PET on other samples using different DNA damaging agents such as IR or γ-radiations may show how these studies are different in dissecting p53 signaling. However, ChIP-PET might prove to be an important development both in cancer genomics and in genome biology in general.

2.3. Serial Analysis of Binding Elements

SABE is another genomic technology for globally identifying binding sites of mammalian transcription factors in vivo. It involved subtractive hybridization of ChIP-enriched DNA fragments followed by the generation and analysis of concatamerized sequence tags. Recently, this approach was applied to search for p53 target genes in the human genome *(26)*. The authors have identified several previously described p53 targets in addition to numerous potentially novel targets, including the DNA mismatch repair genes *MLH1* (mutL homolog 1) and *PMS2* (postmeiotic segregation increased 2). Both of these genes were determined to be responsive to DNA damage and p53 activation in normal human fibroblasts and have p53-response elements within their first intron. These two genes may serve as a sensor in DNA repair mechanisms and a critical determinant for the decision between cell-cycle arrest and apoptosis. The identification of MLH1 and PMS2 as direct targets for p53 defines a signaling pathway that couples two important cellular guardian pathways, growth arrest, and apoptosis. This approach is similar in concept to the genome-wide mapping technique described by Roh et al. *(27)*, which was used to localize hyperacetylated histone H3 protein on the *Saccharomyces cerevisiae* genome.

2.4. Serial Analysis of Gene Expression

Immunoprecipitated DNA fragments from ChIP experiments can be cloned and sequenced. However, because a significant amount of background DNA will still be present in the immunoprecipitated DNA material, it is difficult to distinguish between genuine binding sites and noise without further molecular validation. In order to enhance the coverage of the genuine binding sites, SAGE is another sequencing strategy, which was originally developed for counting transcripts and was also recently applied to genome scanning for transcription-factor binding sites and histone modification.

SAGE is based on high-throughput sequencing of concatamerized tags derived from target DNA enriched by ChIP. SAGE was first used in yeast to confirm that RNA polymerase III genes are the most prominent targets of the TATA-box binding protein. Then it was used to identify several previously unknown binding targets of human transcription factor E2F4 that was independently validated by promoter-specific PCR and microarray hybridization. SAGE provides a means of identifying the chromosomal targets of DNA-associated proteins in any sequenced genome *(28)*. SAGE experiments were also carried out to complement proteomics data. Although it is not possible to deduce quantitative protein expression profiles from mRNA analysis, SAGE provides an overview of gene expression, which can be compared qualitatively to the proteomics result. Recently, this approach was described by de Souza et al. 2006 *(29)*. These authors reported the differential expression of genes and the differential accumulation of proteins in murine melanoma model. The reactive oxygen species was validated to cause DNA damage and to trigger p53-dependent apoptosis.

2.5. RNA Interference Knockdown

The RNAi technology was also used to identify p53 target genes *(30)*. The authors used RNAi to identify new genes that regulate apoptosis in *Caenorhabditis elegans* germ line. They reported that germ line apoptosis depends on the caspase CED-3 and that the expression of *CEP-1* gene (a p53 homolog) is blocked by Bcl-2-like protein, CED-9, and by p21-like checkpoint gene, HUS-1. The study further indicated that the loss of function of both CEP-1 and HUS-1 would completely suppress the germ line apoptosis phenotype of all but five genes, suggesting that p53 (CEP-1) may be a key sensor of germ line cell stress in *C. elegans* and many RNAi candidates indirectly affect germ cell apoptosis through the activation of quality control checkpoint pathways. In many cases, RNAi knockdown approaches do not faithfully reproduce the known loss of function phenotypes; thus, genetic mutations should be employed to confirm the RNAi results. RNAi screens have been used to determine new gene functions and subsequently could be used to provide useful genetic entry points into the study of cellular mechanisms that control the cell survival, genomic stability, and genetic network that regulates p53 signaling.

3. PROTEOMICS AND P53 TARGET IDENTIFICATION

Proteomics is an emerging discipline in the post-genomic era of biomedical research. While genomic approaches have been utilized to establish the "blue-prints" of p53 signaling and its target genes, proteomics has become a necessary tool to validate such information. Although the data obtained from functional genomics may explain more about p53 signaling, several other mechanisms involved are not gene mediated. In addition, information obtained from such study is limited, especially when post-transcriptional and post-translational modifications occur. The level of p53 can be modified by several different products or proteins, which directly determine differential cellular functions and responses. Therefore, integration of genomics and proteomics information with respect to one gene is important to develop gene(s)-specific testable hypothesis, as outlined in Fig. 3.

To date, there have been only few studies that applied the proteomic approach to p53 signaling, and all of them dealt mainly with inducible systems to identify the downstream targets. By comparing the proteome of human colorectal cancer cells transfected with inducible p53 (DLD-1.p53) with that of the control DLD-1 cell line using amino acid-coded mass tagging (AACT)-assisted mass spectrometry (MS), Gu et al. *(31)* identified proteins that are up-regulated at the execution stage of the p53-mediated apoptosis. Differentially expressed proteins induced by p53 overexpression were quantitated by

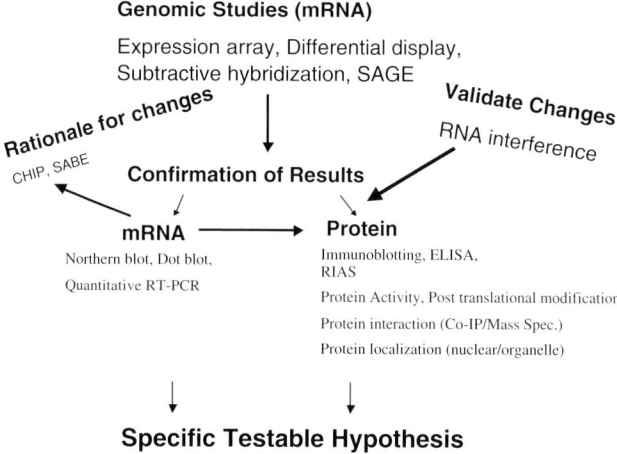

Fig. 3. Flow diagram for generating a specific testable hypothesis using genomics and proteomics approaches. Arrows indicate the sequence of events to reach a testable hypothesis in addition to the literature search. The techniques used are listed in normal letters. ChIP, chromatin immunoprecipitation; Co-IP, Co-immunoprecipitation; ELISA, enzyme linked immunosorbent assay; mRNA, messenger RNA; RIA, radioimmunoassay; RT-PCR, reverse transcription followed by polymerase chain reaction; SABE, serial analysis of binding elements; SAGE, serial analysis of gene expression.

comparing the intensity of an AACT peptide with that of its unlabeled counterpart. Using complementary AACT-LC-MS/MS and matrix-assisted laser desorption ionization-time-of-flight mass spectrometry (MALDI-TOF MS) approaches, the authors found 81 proteins that were up-regulated and 23 proteins that were down-regulated in response to p53 overexpression. The regulated proteins were associated with cell cycle arrest, p53-binding protein chaperones, plasma membrane dynamics, stress response, antioxidant proteins, and glycolysis and ATP generation/transport. Thus, the authors of this study were able to identify many proteins that are involved in multiple pathways controlled by p53.

Mutations of p53 are the most common genetic alterations found in human tumors. Most pathogenetic modifications are missense mutations that abolish the p53 DNA-binding function. To study the effect of these mutations on p53 regulation of cell function such as transformation, Paron and his group *(32)* analyzed the proteomes of the rat thyroid epithelial cell line PC CI3, its derivative PC CI3 V143A and PC CI3 S392A. Introduction of the V143A mutant p53 allele, which abolishes the p53 DNA-binding function, modified the expression of 23.6% of total proteins, while that of PC CI3 S392A mutant cells with mutation located outside the DNA-binding domain showed changes only in 14.0% of the total protein components. Thus, these authors were able to show that mutations of p53 DNA-binding domain have more impact on p53 signaling than that located outside the DNA-binding domain.

It is well known that proteomics analysis using two-dimensional polyacrylamide gel electrophoresis has some limitations including the labor-intensive nature of the technique, its low sensitivity, and a significant bias against proteins with extreme molecular weights or isoelectric points. Thus, isotopic-coded affinity tag reagents and high-throughput MS were used to study p53-dependent cell death processes that occur in the primary mouse wild-type (p53+/+) cortical neurons after DNA damage response with camptothecin, a topoisomerase I inhibitor *(33)*. A total of 191 peptides corresponding to 150 proteins prepared from control and camptothecin-treated wild-type postnatal mouse cortical neurons were identified and quantitated. The results indicated that 65% of proteins derived from camptothecin-treated neurons showed no significant change in their abundance levels. Several proteins involved in energy production and oxidative stress were up-regulated after DNA damage. However, DNA damage caused a p53-dependent decrease in the expression of many members of the protein kinase A signaling pathway. This could be due to the shutdown of the cell's transcriptional machinery by camptothecin.

4. INTEGRATION OF PROTEOMICS IN P53 SIGNALING

The most striking observation of our genomics study *(3)* is that *S100A4* gene has the most intensely p53-dependent expression (>25-fold) as shown in the E region of the GEM (Fig. 2). Differential expression of S100A was confirmed by immunoblots, which demonstrated a strong signal in the p53+/+ cells and no detectable signal in p53–/– cells. Others have also supported these observations *(34)*. These findings suggest a close association between S100A4 expression and p53 function. S100A4/mst1 is a M_r 11,000 protein that belongs to the S100 family of Ca^{2+}-binding proteins, different members of which have diverse cellular functions *(35)*. Calcium binding induces conformational changes in the S100 protein structure, allowing interaction with target proteins including p53 *(21,36)*. Although no p53-binding sites were identified in the promoter of the S100A4 gene *(25)*, the expression of S100A4 protein is abundant in cells with wild-type p53. Additionally, S100A4 protein binds to the tetramerization domain of p53 that also contains PKC phosphorylation site located at the C terminus of p53 protein *(21)* resulting in the inhibition of p53 phosphorylation, although the interaction between p53 and S100A4 may result in the promotion of tumor metastasis *(37)*. Thus, S100A4 might directly contribute in p53 signaling. The extent to which reduced expression of S100A in p53–/– cells would alter the interaction of S100A4 with other interacting proteins was further investigated. For this purpose, we used functional proteomics approach. In this study, we performed immunoprecipitation assays on cell lysates from control and topotecan-treated p53+/+ and p53–/– HCT116 cells using antibodies against S100A4 protein. The immunoprecipitated proteins from each cell lysates were subjected to denaturing gel electrophoresis, and differentially expressed proteins in each cell line were subjected to mass fingerprinting with MALDI-TOF MS, as we have previously reported *(38)*. Figure 4 shows a Coomassie blue-stained gel of protein co-immunoprecipitated with S100A4 antibody from each cell line following treatment with topotecan. It is clearly observed

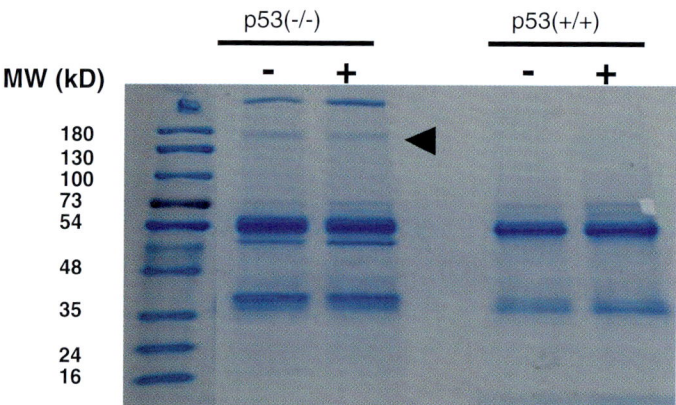

Fig. 4. Coomassie blue-stained gel of proteins co-immunoprecipitated with S100A4 monoclonal antibody from isogenic colorectal cancer cells. Two independent co-immunoprecipitated samples from untreated control (–) and cells treated with 1 µM topotecan (+) were loaded. The gels were stained with Coomassie blue. Molecular masses of protein size markers are indicated (kDa). The arrowhead indicates the band of stained proteins excised for enzymatic digestion by trypsin and subsequent mass fingerprinting with matrix-assisted laser desorption ionization-time-of-flight mass spectrometry.

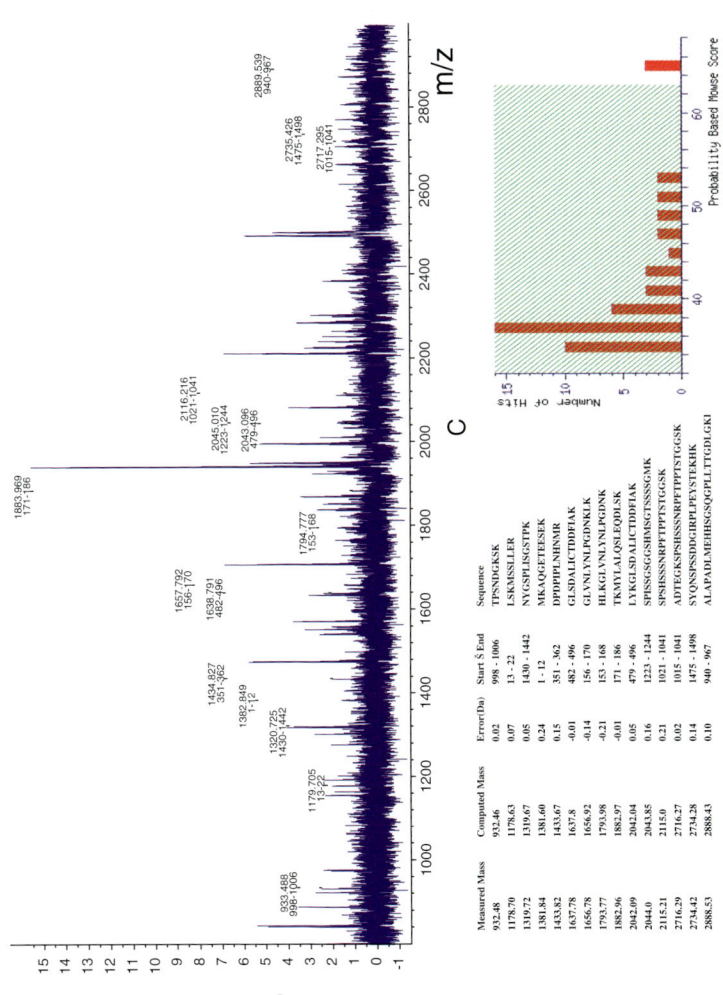

Fig. 5. matrix-assisted laser desorption ionization-time-of-flight (MALDI-TOF) mass spectrometric analysis of RB18A protein. (A) Peptide mass fingerprinting of in-gel tryptic digest of 180-kDa bands generated by MALDI-TOF mass spectrometer. (B) Identified sequence of RB18A protein. Sequence coverage of the matched protein was 13.2%. (C) MOSCOT search of identified sequence. The probability score of the protein candidate (Klenow fragment) was 64. Protein scores greater than 61 are significant (p < 0.05).

in Fig. 4 that there are three *(3)*distinct bands detectable in p53–/– cells, whereas these bands are completely absent in p53+/+ cells. The three bands were excised and subjected to in-gel digestion by trypsin. Figure 5A illustrates the MALDI mass spectrum acquired from the peptide mixture resultant from in-gel digestion of the 180-kDa band marked with an arrow head, as shown in Fig. 4. On average, 15 peptide masses were identified in this spectrum (Fig. 5A) that agreed with the expected peptide masses within a mass tolerance of window of +/– 0.24 Daltons. The probability-based MOWSE score resultant from the MOSCOT database search (http://www.matrixscience.com) for this 180-kDa band indicated RB18A was the only protein that was identified within the scoring region where the probability of a match has less than a 5% chance of being a random event (Fig. 5C). The protein coverage of peptides for RB18A covered 13.2% of the protein sequence (Fig. 5B). Furthermore, these results were repeated with multiple co-immunoprecipitation (co-IP) experiments that all resulted in the identification of RB18A as a new protein-binding partner to S100A4, especially in HCT-116 (p53–/–) cells. RB18A protein is recognized by various monoclonal antibodies to p53 and shares identical properties with p53 such as DNA binding and tetramerization, despite no significant homologies with p53 protein at the sequence level *(39)*. The interaction of S100A4 protein with RB18A was confirmed by reversed co-IP using antibodies to TRAP-220, a protein that is almost identical to RB18A, as shown in Fig. 6. RB18A

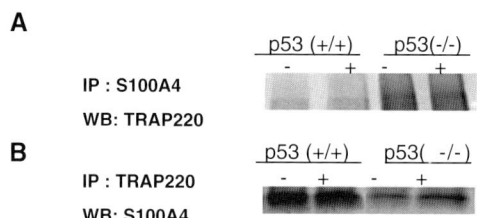

Fig. 6. Confirmation of protein interactions between S100A4 and RB18A in isogenic colorectal cells. The cell lysates (500 µg) from both p53+/+ and p53–/– cells untreated (–) and cells treated with 1 µM topotecan (+) were immunoprecipitated separately with 1 µg of primary antibodies specific to S100A4 (A) and TRAP-220 (B). The immunoprecipitated proteins were subjected to immunoblotting with antibodies to interacting partner after separating by denaturing gel electrophoresis (4–15% gradient gel). The primary antibody was detected with species-specific secondary antibody conjugated to IR dyes using LI-COR odyssey™ detection system.

shares 98% nucleotide sequence identity and functional properties with TRAP-220 and DRIP-205 *(40)*. It was shown that both TRAP-220 and DRIP-205 interact with distinct hormone-activated nuclear receptor. RB18A induces the expression of Mdm-2 protein by binding to Mdm-2 promoter resulting in decreased levels of p53 wild-type protein due to Mdm-2-mediated proteasomal degradation of p53 *(41)*. However, RB18A binds with mutant p53 and inhibits Mdm-2 promoter transactivation resulting in increased expression of mutant p53. Based on these functional roles of RB18A and the fact that its interaction with p53 involves Mdm-2, we also observed that Mdm-2 transcript expression was elevated in p53–/– cells *(3)*. These observations were confirmed by immunoblotting analysis shown in Fig. 7. In light of these results and the fact that Mdm-2, p53, and RB18A are proteins that could be localized to cell nucleus, it is plausible to explore further the role

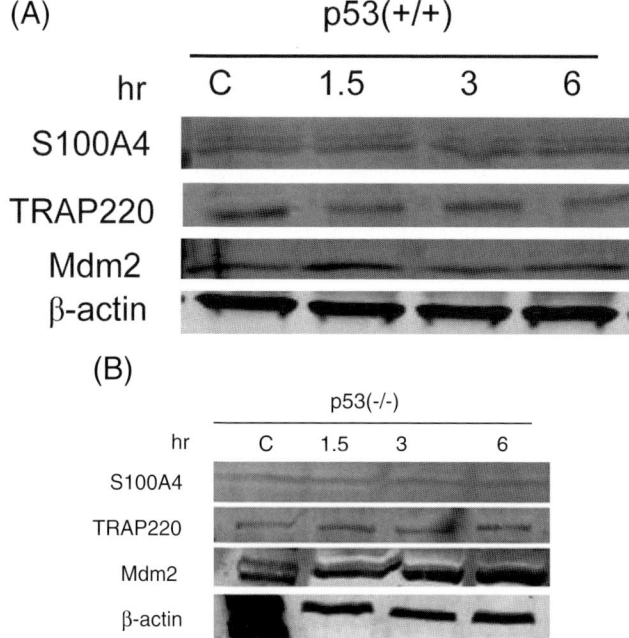

Fig. 7. Expression of S100A4, RB18A, and Mdm-2 proteins in isogenic colorectal cancer cells. The cell lysates (50 µg) from both (A) p53(+/+) and (B) p53(–/–) cells untreated and treated with 1 µM topotecan for 1.5, 3, and 6 h were subjected to denaturing gel electrophoresis (4–15% gradient gel) followed by immunoblotting with antibodies to the indicated protein. The primary antibody was detected by species-specific secondary antibody conjugated to IR dyes as described in Fig. 6.

of S100A4 protein and its interaction with RB18A protein in p53–/– cells. As p53–/– cells express higher level of Mdm-2 transcript and protein compared to p53+/+ cells, we can certainly test the hypothesis that novel interaction of S100A4 protein with RB18A in the nucleus results in induced expression of Mdm2 in p53–/– cells. Therefore, the use of genomic results followed by proteomic studies could be very well provide a paradigm in the molecular regulation of p53 protein signaling.

5. CONCLUSIONS

Advances in technology and methodology of high-throughput experimental approaches, such as genomics, proteomics, and bioinformatics, provide an opportunity to dissect the regulatory aspects of many tumor suppressor proteins such as p53. These technologies provide a limited catalog of p53 target genes, but the application of additional experimental approaches, including the use of model organisms, may eventually allow the construction of a complete "p53 molecular interaction map." The ability to construct such a map is already contributing to our understanding of certain aspects of the basic biology of p53. This integrated understanding of p53 signaling will allow more complete characterization of the role of p53 as a "guardian of the genome." The identification as well as the catalogs of p53 target genes is the first step toward the establishment of the p53 molecular interaction map. Genomics and proteomics studies involving genes such as S100A4 will provide new information on genes that lack p53-binding sites in their promoters and yet are abundantly expressed in p53-containing cells. This would generate a new hypothesis to identify unique sets of transcription factors that modulate S100A4 expression. Our integrated approach of S100A4 and p53 proved that instead of focusing only on genes or proteins that are regulated by p53, applying an integrative approach seems to be the ideal way for future understanding of the role of tumor suppressors such as p53 and its presently unknown downstream mediators, which eventually will contribute to our ability for developing better selective therapy for tumor metastasis of many cancers.

ACKNOWLEDGMENT

We apologize for not being able to cite all relevance works due to limitations of space. We thank the Proteomics Core Facility at Washington State University for technical assistance. HCT-116 and an isogenic p53 knockout cell line were kindly provided by

Dr. Bert Vogelstein (The John Hopkins Kimmel Cancer Center, Baltimore, MD). The work is supported by grants from the US Army for Breast Cancer Research and the S.G. Komen Foundation for Breast Cancer Research.

REFERENCES

1. Lander ES, Linton LM, Birren B, Nusbaum C, Zody MC, Baldwin J, Devon K, Dewar K, Doyle M, FitzHug W, et al. Initial sequencing and analysis of the human genome. *Nature* 2001; 409: 860–921.
2. Venter JC, Adams MD, Myers EW, Li PW, Mural RJ, Sutton GG, Smith HO, Yandell M, Evans CA, Holt RA, et al. The sequence of the human genome. *Science* 2001; 291; 1304–1351.
3. Daoud SS, Munson PJ, Reinhold W, Young L, Prabhu VV, Yu Q, LaRose J, Kohn KK, Weinstein JN, Pommier Y. Impact of p53 knockout and topotecan treatment on gene expression profiles in human colon carcinoma cells: A pharmacogenomic study. *Cancer Res* 2003; 63: 2782–2793.
4. Scherf U, Ross DT, Waltham M, Smith LH, Lee JK, Tanabe L, Kohn KW, Reinhold WC, Myers TG, Andrews DT, Scudiero DA, Eisen MB, Sausville EA, Pommier Y, Botstein D, Brown PO, Weinstein JN. A gene expression database for the molecular pharmacology of cancer. *Nat Genet* 2000; 24: 236–244.
5. Zeeberg BR, Qin H, Narasimhan S, Sunshine M, Cao H, Kane DW, Reimers M, Stephens RM, Bryant D, Burt SK, Elnekave E, Mari DM, Wynn TA, Cunningham-Rundles C, Stewart DM, Nelson D, Weinstein JN. High-throughput GoMiner, an 'industrial-strength' integrative gene ontology tool for interpretation of multiple-microarray experiments, with application to studies of common variable immune deficiency (CVID). *BMC Bioinformatics* 2005; 6: 168–186.
6. Kohn KW, Aladjem MI, Weinstein JN, Pommier Y. Molecular interaction maps of bioregulatory networks: A general rubric for systems biology. *Mol Biol Cell* 2006; 17: 1–13.
7. Weinstein JN. 'Omic' and hypothesis-driven research in the molecular pharmacology of cancer. *Curr Opin Pharmacol* 2002; 2: 361–365.
8. Abramovitz M, Leyland-Jones B. A systems approach to clinical oncology: Focus on breast cancer. *Proteome Sci* 2006; 4: 5–20.
9. Mustacchi R, Hohmann S, Nielsen J. Yeast systems biology to unravel the network of life. *Yeast* 2006; 23: 227–238.
10. Aloy P, Russell RB. Structural systems biology: Modeling protein interactions. *Nat Rev Mol Cell Biol* 2006; 7: 188–197.
11. Caspi R, Foerster H, Fulcher CA, Hopkinson R, Ingraham J, Kaipa P, Krummenacker M, Paley S, Pick J, Rhee SY, Tissier C, Zhang P, Karp PD. MetaCyc: A multiorganism database of metabolic pathways and enzyme. *Nucleic Acids Res* 2006, 34 (Database issue): D511–D516.
12. Levine, AJ. p53, the cellular gatekeeper for growth and division. *Cell* 1997; 88: 323–331.
13. Vogelstein B, Lane D, Levine AJ. Surfing the p53 network. *Nature (Lond.)* 2000; 408: 307–310.
14. Zambetti G, Bargonetti J, Walker J, Pives C, Levine A. Wild-type p53 mediates positive regulation of gene expression through a specific DNA sequence element. *Genes Dev* 1992; 6: 1143–1152.

15. Courtois S, de Fromentel CC, Hainaut P. p53 protein variants: structural and functional similarities with p63 and p73 isoforms. *Oncogene* 2004; 23: 631–638.
16. Yee KS, Vousden KH. Complicating the complexity of p53. *Carcinogenesis* 2005; 26: 1317–1322.
17. Bourdon JC, Fernandes K, Murray-Zmijewski F, Liu G, Diot A, Xirodimas DP, Saville MK, Lane, DP. p53 isoforms can regulate p53 transcriptional activity. *Genes Dev* 2005; 19: 2122–2137.
18. Zhao R, Gish K, Murphy M, Yin Y, Notterman D, Hoffman WH, Tom E, Mack DH, Levine AJ. Analysis of p53-regulated gene expression patterns using oligonucleotide arrays. *Genes Dev* 2000; 14: 981–993.
19. Kannan K, Amariglio N, Rechavi G, Jakob-Hirsch J, Kela I, Kaminski N, Getz G, Domany E, Givol D. DNA microarrays identification of primary and secondary target genes regulated by p53. *Oncogene* 2001; 20: 2225–2234.
20. Kannan K, Kaminski N, Rechavi G, Jakob-Hirsch J, Amariglio N, Givol D. DNA microarray analysis of genes involved in p53 mediated apoptosis: activation of Apaf-1. *Oncogene* 2001; 20: 3449–3455.
21. Grigorian M, Andresen S, Tulchinsky E, Kriajevska M, Carlberg C, Kruse C, Cohn M, Ambartsumian N, Christensen A, Selivanova G, Lukanidin E. Tumor suppressor p53 protein is a new target for the metastasis-associated Mts1/S100A4 protein. *J Biol Chem* 2001; 276: 22699–22708.
22. Ren B, Robert F, Wyrick JJ, Aparicio O, Jennings EG, Simon I, Zeitlinger J, Schreiber J, Hannett N, Kanin E, et al. Genome wide location and function of DNA binding proteins. *Science* 2000; 290: 2306–2309.
23. Weinmann AS, Bartley SM, Zhang T, Zhang MQ, Farnham PJ. Use of chromatin immunoprecipitation to clone novel E2F target promoters. *Mol Cell Biol* 2001; 21: 6820–6832.
24. Hug BA, Ahmed N, Robbins JA, Lazar MA. A chromatin immunoprecipitation screen reveals protein kinase cβ as a direct RUNX1 target gene. *J Biol Chem* 2004; 279: 825–830.
25. Wei C-L, Wu Q, Vega VB, Chiu KP, Ng P, Zhang T, Shahab A, Yong HC, Fu Y, Weng Z, Liu J, Zhao XD, Chew J-L, Lee YL, Kuznetsov VA, Sung W-K, Miller LD, Lim B, Liu ET, Yu Q, Ng H-H, Ruan Y. A global map of p53 transcription-factor binding sites in the human genome. *Cell* 2006; 124: 207–219.
26. Chen J, Sadowski I. Identification of the mismatch repair genes PMS2 and MLH1 as p53 target genes by using serial analysis of binding elements. *Proc Natl Acad Sci USA* 2005; 102: 4813–4818.
27. Roh TY, Ngau WC, Cui K, Landsman D, Zhao K. High-resolution genome-wide mapping of histone modifications. *Nat Biotechnol* 2004; 22: 1013–1016.
28. Kim J, Bhinge AA, Morgan XC, Iyer VR. Mapping DNA-protein interactions in large genomes by sequence tag analysis of genomic enrichment. *Nat Methods* 2005; 2: 47–53.
29. de Souza GA, Godoy LM, Teixeira VR, Otake AH, Sabino A, Rosa JC, Dinarte AR, Pinheiro DG, Silva WA, Eberlin MN, Chammas R, Greene LJ. Proteomic and SAGE profiling of murine melanoma progression indicates the reduction of proteins responsible for ROS degradation. *Proteomics* 2006; 6: 1460–1470.
30. Lettre C, Kritikou EA, Jaeggi M, Calixto A, Fraser AG, Kamath RS, Ahringer J, Hengartner MO. Genome-wide RNAi identifies p53-dependent and independent regulators of germ cell apoptosis in C. elegans. *Cell Death Differ* 2004; 11: 1198–1203.

31. Gu S, Liu Z, Pan S, Jiang Z, Lu H, Amit O, Bradbury EM, Hu CA, Chen X. Global investigation of p53-induced apoptosis through quantitative proteomic profiling using comparative amino acid-coded tagging. *Mol Cell Proteomics* 2004; 3: 998–1008.
32. Paron I, D'Ambrosio C, Scaloni A, Berlingieri MT, Pallante PL, Fusco A, Bivi N, Tell G, Damante G. A differential proteomic approach to identify proteins associated with thyroid cell transformation. *J Mol Endocrinol* 2005; 34: 199–207.
33. Yu LR, Conrads TP, Uo T, Issaq HJ, Morrison RS, Veenstra TD. Evaluation of the acid-cleavable isotope-coded affinity tag reagents: Application to camptothecin-treated cortical neurons. *J Proteome Res* 2004; 3: 469–477.
34. Mann K, Hainaut P. Aminothiol WR1065 induces differential gene expression in the presence of wild-type p53. *Oncogene* 2005; 24: 3964–3975.
35. Donato, R. S100: A multigenic family of calcium-modulated proteins of the EF-hand type with intracellular and extracellular functional roles. *Int J Biochem Cell Biol* 2001; 33: 637–668.
36. Scotto C, Deloulme JC, Rousseau D, Chambaz E, Baudier J. Calcium and S100B regulation of p53-dependent cell growth arrest and apoptosis. *Mol Cell Biol* 1998; 18: 4272–4281.
37. Garrett SC, Varney KM, Weber DJ, Bresnic AR. S100A4, a mediator of metastasis. *J Biol Chem* 2006; 281: 677–680.
38. Rehman A, Chahal MS, Tang X, Bruce JE, Pommier Y, Daoud SS. Proteomic identification of heat shock protein 90 as a candidate target for p53 mutation reactivation by PRIMA-1 in breast cancer cells. *Breast Cancer Res* 2005; 7: R765–R774.
39. Drane P, Barel M, Balbo M, Frade R. Identification of RB18A, a 205 kDa new p53 regulatory protein which shares antigenic and functional properties with p53. *Oncogene* 1997; 15: 3013–3024.
40. Lottin-Divoux S, Barel M, Frade R. RB18A enhances expression of mutant p53 protein in human cells. *FEBS Lett* 2005; 579: 2323–2326.
41. Frade R, Balbo M, Barel M. RB18A regulates p53-dependent apoptosis. *Oncogene* 2002; 21: 861–866.

3 Proteomic Profiling of Tyrosine Kinases as Pharmacological Endpoints for Targeted Cancer Therapy

Moulay A. Alaoui-Jamali and Devanand Pinto

CONTENTS

1. INTRODUCTION
2. PROTEIN TYROSINE KINASES: VERSATILE TARGETS FOR HUMAN CANCER
3. GENOMIC AND STRUCTURAL FEATURES OF PROTEIN KINASES IN CANCER: DIVERSE MUTATIONAL PATTERNS AND IMPACT ON TARGETED THERAPY
4. PROTEOMIC PROFILING OF TYROSINE KINASES

SUMMARY

Protein tyrosine kinases (PTKs) and their substrates are emerging as attractive therapeutic targets and potential biomarkers for molecular classifications, prediction of clinical outcome and monitoring response to cancer treatments. The exciting move toward kinase-targeted therapy has brought new technical challenges in profiling protein kinase genes and proteins as surrogate clinical biomarkers, particularly in light of clinical data correlating specific mutations in PTKs with either sensitivity or resistance to targeted therapy. This chapter discusses the impact of mutations on protein conformation, protein phosphorylation, and drug response and the utility of proteomic technology to mine the phosphoproteome for PTK profiling and prediction of response to targeted therapies.

Key Words: Cancer; protein tyrosine kinases; targeted therapy; proteomics

1. INTRODUCTION

Protein kinases, including protein tyrosine kinases (PTKs) and serine/threonine kinases, are one of the largest classes of proteins and are implicated in a number of human diseases, including cancer, leukemia, diabetes, and congenital syndromes (summarized in http://www.sdsc.edu/Kinases/). With the completion of the human genome project, the protein kinase family has expanded to include 20 subfamilies of over 90 tyrosine kinases, 10 subfamilies of over 32 non-receptor kinases, and over 2000 protein serine/threonine kinases *(1)* (Fig. 1).

In addition to being attractive targets for drug development, PTKs and their substrates are emerging as potential biomarkers with invaluable applications for molecular classifications, prediction of clinical outcome and monitoring response to treatment. Several successful agents that target protein kinases, particularly in the context of cancer therapeutics, have reached clinical trials or have been clinically approved. These include herceptin (trastuzumab) which targets the ErbB-2/Her-2 receptor for metastatic breast cancer, Gleevec (imatinib mesylate) which targets Bcr-Abl for chronic myelogenous leukemia (CML) and gastrointestinal stroma tumors, iressa (gefitinib) for chemotherapy-recalcifiant non-small cell lung carcinoma, avastin which targets vascular endothelial growth factor (VEGF) for colorectal cancer, VEGF trap for non-Hodgkins lymphoma, and others (reviewed in ref. *2*). With this exciting move toward kinase-targeted therapy comes new technical challenges in profiling protein kinase genes and proteins as surrogate clinical biomarkers. This is of particular interest in light of recent clinical data showing that specific mutations in PTK (Section 3) can predict either sensitivity or resistance to a specific targeted therapy.

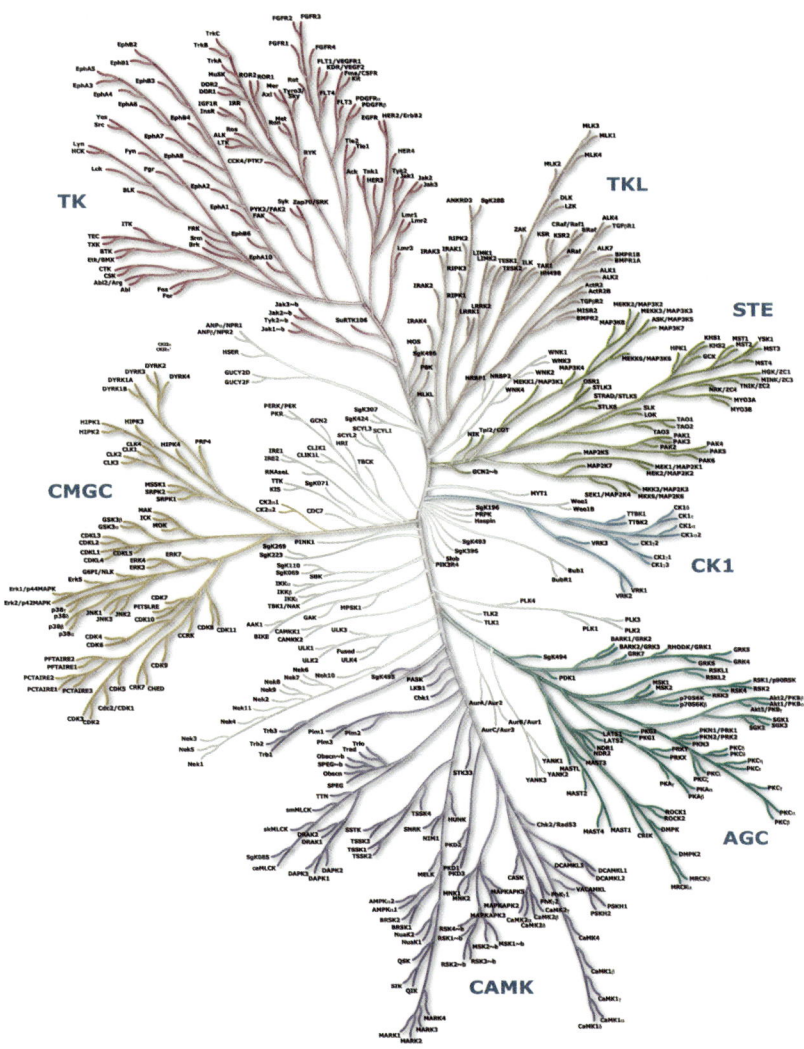

Fig. 1. A phylogenetic analysis of the human protein kinase families from the identified human kinome. The dendrogram was kindly revised by Dr. Michael Melnick and Dr. Dave Comb and reproduced with permission from Cell Signaling Technology Inc. (Danvers, MA). CAMK, calcium/calmodulin-dependent protein kinases; CKI, cyclin-dependent kinase inhibitors; STE, serine/threonine kinases; TK, tyrosine kinases; TKL, tyrosine kinase-like group. AGC family of protein kinases comprise cAMP-dependent protein kinase (PKA), Protein kinase G, and Protein kinase C. The GMGC kinases comprise cyclin-dependent kinases (CDKs), mitogen-activated protein kinases (MAPKs), glycogen synthase kinases (GSKs), and CDK-like kinases(CLKs), which are collectively termed the CMGC group.

This chapter interfaces current knowledge on PTKs as therapeutic targets and the utility of proteomic profiling as a tool to screen for surrogate pharmacological biomarkers to predict responses and relapses. The review is not meant to be comprehensive; aspects related to tyrosine kinase signaling, targeting, ongoing clinical trials with kinase inhibitors, and detailed proteomic instrumentation are not dealt with extensively in this review. However, we refer the reviewer to excellent reviews that provide coverage of these topics and to the accompanying chapters.

2. PROTEIN TYROSINE KINASES: VERSATILE TARGETS FOR HUMAN CANCER

2.1. Tyrosine Kinase Receptor Activation

PTK receptors contain highly conserved catalytic domains with unique subdomain motifs such as adapter or signaling domains [e.g., Src homology (SH)2, SH3, or PH domains] that distinguish them from similar catalytic domains found in serine/threonine and dual-specificity kinases *(3)*. In addition to the catalytic domain, PTK receptors contain a transmembrane domain and an extracellular binding domain. In some of the non-receptor PTKs, such as Src, the protein contains a myristoylation site at the N-terminal domain of the protein that is required for membrane localization.

The first step in PTK receptor activation is the interaction of the receptor with its growth factor ligand leading to receptor phosphorylation. Although receptor ligands are diverse and many can induce a similar level of overall receptor phosphorylation, the specific tyrosine residues phosphorylated within each receptor can differ with each growth factor, which contributes to signaling specificity. In addition, interaction between the ligand and the receptor can be modulated independently by unrelated receptors and factors *(4)*.

Ligand-induced homo- and hetero-dimerization is a recurring and less understood theme in PTK receptor function. Some growth factors such as the Epidermal Growth Factor (EGF)-like ligand are suggested to be bivalent by virtue of the existence of two separate regions, one that binds to a primary ErbB receptor with high affinity and narrow specificity and the other binds to a dimerizing receptor with lower affinity and broader specificity (reviewed in refs *5,6*). Other growth factors such as platelet-derived growth factor, a member of cysteine knot-containing proteins, are dimeric and their receptor encodes a single ligand-binding domain. As a result, two receptors bind to each monomer of the ligand resulting in dimerization of the receptors. Growth factors such as the fibroblast growth factor (FGF) has two binding sites for the receptor within a single molecule of FGF, and each receptor molecule also has two binding sites for the ligand *(7–9)*. As a result, the ligand–receptor complex consists of two molecules of FGF; the high affinity binding of FGF to its receptor requires heparan sulfate proteoglycans, which spatially organize the ligands to functionally associate with the receptor. In contrast to the above examples, the insulin receptor is a dimer even before the binding of monomeric insulin. Another complex and controversial situation that is gaining wider acceptance is the ligand-induced receptor aggregation and the existence of preformed receptor aggregates. For instance, formation of tetramers or high order oligomers has been suggested from molecular modeling studies of the kinase domain of EGFR and ErbB-2 *(10)*.

2.2. Receptor Phosphorylation

In general, activation of a PTK receptor involves the reconfiguration of various domains, including the activation loop and the orientation of upper and lower lobes of the kinase *(11)*. Tyrosine kinases are often phosphorylated in the activation loops contributing to the conformational changes that lead to activation of the kinase. Ligand-induced dimerization can bring the kinase domains into close proximity and facilitates transphosphorylation at the activation loop tyrosines *(12,13)*. Each receptor subunit within the dimeric complex in turn cross-phosphorylates tyrosine residues in the activation loop (A-loop) region of the kinase domain of its neighbor, removing a physical constraint and enhancing kinase activity. Receptor subunits then cross-phosphorylate each other on specific tyrosine residues responsible for recruitment and activation of signaling molecules. This process is crucial for receptor transactivation and phosphorylation, particularly for orphan receptors such as ErbB-2, which can be activated by transphosphorylation by other members of this receptor family.

Receptor phosphorylation involves direct transfer of γ-phosphate from ATP to tyrosyl 4´-hydroxyls, resulting in the creation of phospho-tyrosyl residues within the substrate proteins. Once formed, phospho-tyrosine residues provide critical functionality needed for recognition/association by pTyr-binding molecules that lead to the formation of multiprotein complexes responsible for further signal propagation. Among these are SH2 domains and phosphotyrosyl-binding (PTB) domains.

Tyrosine kinase phosphorylation is modulated by tyrosine phosphatases, which have the capability to functionally terminate the kinase activity by dephosphorylating tyrosine residues involved in catalytic function. Several protein tyrosine phosphatases (PTPs) are capable of dephosphorylating tyrosine by removing pTyr phosphoryl groups and returning tyrosyl residues to their nonphosphorylated state. PTPs often function as downregulators of PTK-depending pathways, but they can also serve in positive signaling roles by neutralizing inhibitory pTyr residues *(14)*; the inappropriate activation of pTyr residues has been associated with many diseases, including cancer *(15)*. The prevelance of mutations in PTPs can be quite high, such as in the case of colorectal cancer where over 50% of tumors carry mutations in PTPs *(16)*. In addition to phosphatases, non-phosphatase proteins may contribute to the prevention of PTK receptor activation. An interesting example is the transmembrane protein Kek that has been shown to interact with the Drosophila EGFR and antagonize its activity *(17)*.

2.3. Tyrosine Kinase Signaling Diversity

PTK receptors are coupled to multiple signal transduction cascades that are regulated by a multitude of feedback loops and a high level of cross-talks between distinct pathways shared by many receptors from different families. This diversity of signaling networks create alternative compensatory mechanisms to overcome therapeutic activity of targeted agents as well as cytotoxics. In general, receptor activation and tyrosine phosphorylation lead to recruitment of multiple specific signaling molecules *(18–20)*. The Ras, PI-3-K, PKC, JAK-STAT, and cytokine pathways are well characterized. Many of these signaling molecules are increasingly appreciated as a family of enzymes instead of single entities and mutations in some of these molecules occur in cancer, adding to the complexity of PTK pathways both in size and in scope. An example of such complexity is provided by a study by Schulze and colleagues *(21)* who reported an exhaustive computational mapping of molecular interactions involving ErbB phosphotyrosine kinases. Each of these interactions can provide additional framework and insight into the fundamental need of using protein array and phosphoproteomic profiling as a tool to study surrogate pharmacological biomarkers for cancer therapeutics.

3. GENOMIC AND STRUCTURAL FEATURES OF PROTEIN KINASES IN CANCER: DIVERSE MUTATIONAL PATTERNS AND IMPACT ON TARGETED THERAPY

Cancer cells accumulate assorted mutations in many tyrosine kinase receptors *(22)*; useful information on somatic mutations in PTK is found in http://www.sanger.ac.uk/cosmicin), often resulting in deregulation of signals responsible for transducing extracellular signals into modulation of gene transcription. Most mutations render a specific kinase pathway overactive while very few can lead to the inhibition of kinase activity. As a consequence, deregulated PTK and other kinases represent the hallmark of many cancers, making those placed in key positions in the signaling network attractive targets for drug development. Two relevant examples are provided—Bcr-Abl and c-Kit, and ErbB receptors.

The Bcr-Abl oncogenic kinase results from a translocation between chromosome 9 and 22 that leads to the formation of the constitutively active Bcr-Abl kinase (cytologically expressed as the Philadelphia chromosome). This reciprocal translocation, found in over 90% of CML patients, involves a replacement of the first exon of c-Abl with sequences from the *bcr* gene. The fusion Bcr-Abl kinase activates several pro-survival pathways including Ras-MAPK, JAK-STAT, and PI-3-K-AKT pathways *(23)*.

The discovery of Gleevec (imatinib mesylate), a selective inhibitor of Bcr-Abl kinase and a drug that has greatly changed the management of CML *(24,25)*, augurs well in that the target of this agent is recognized to be important for the maintenance and/or progression of the disease. Therefore, for the drug to be successful, the target must be present. However and as seen with most anticancer drugs, molecular remission with Gleevec is seen in a small proportion of patients (estimated about 5–10% of cases) despite the presence of the target, and even in these cases, it has been postulated that Bcr-Abl cells might be present below the detection level of current laboratory assays used to detect the Bcr-Abl kinase. Relapses may appear when preexisting pool of cells with Bcr-Abl become predominant. Also, over 50% of CML patients with relapses have reactivation of the Bcr-Abl kinase due to mutations in the kinase domain or occasionally gene amplification *(26–29)*. Many of the mutations in the Abl tyrosine kinase found in patients affect the conformation of the kinase domain, preventing binding to Gleevec. Interestingly, it has been reported that these mutations can be present in a small population of cells even before treatment with Gleevec *(30,31)*, and hence, new high resolution and sensitive proteomic/genomic technologies may represent the remedy to detect such a small population of cells. Gleevec also targets Kit kinase that is activated in gastrointestinal stromal tumors (GISTs), where 90% of GISTs carry activating mutations in the *Kit* gene. These mutations are usually in exon 9 or 11 and, interestingly, the best clinical responses are reported in patients who carry mutations in exon 11 *(32,33)*.

In the case of ErbB/Her tyrosine kinases, aberrant expression, amplification, and activation of members of this receptor family is common in many cancers, particularly with ErbB-2. In many instances, this has been correlated with poor prognosis *(34)*. The presence of such a target that is overexpressed and amplified in cancer cells has led to the development of the first humanized antibody for metastatic breast cancer, herceptin/trastuzumab. Herceptin, particularly in combination with chemotherapy, significantly improves the overall response rate, time to progression, duration of response, and survival time compared to chemotherapy alone *(35,36)*. In addition to herceptin, several antibodies and small molecules have been developed to target not only EGFR and ErbB-2 but also multiple ErbB receptors *(2,37)*, based on the assumption that distinct ErbB receptors cooperate through heterodimerization and transphosphorylation for disease progression. As seen with Gleevec, relapses occur in almost all patients who respond to initial anti-ErbB therapy. EGFR gene profiling has identified several deletions and point mutations that can result in increased TK catalytic activity of the receptor. The most prevalent of these mutations is found to be EGFRvIII, an EGFR deletion mutant that lacks exons 2–7, which can arise from gene rearragement or alternative splicing *(38)*. Also, 10% of patients with non-small cell lung cancer (NSCLC) harbor specific mutations in EGFR. Small deletions in amino acids 747–750 of EGFR or point mutations (mostly a replacement of leucine by arginine at codon 858) have been proposed to increase sensitivity of NSCLC to gefitinib (Iressa) via repositioning of critical residues surrounding the ATP-binding cleft of the EGFR tyrosine kinase, thereby stabilizing their interaction with both ATP and its competitive inhibitors *(39,40)*. Patients who have a better overall response to gefitinib in clinical trials have a higher frequency of the responsive mutations than American patients *(40)*. In contrast, resistance to gefitinib is reported to involve at least a point mutation resulting in a replacement of threonine by methionine at codon 790 of EGFR *(41)*. Such somatic mutational profiles are now found to occur in a large number of protein kinase genes *(42)* and could account for resistance or sensitivity to many targeted agents *(43)*.

Most tyrosine kinase inhibitors are ATP competitive in nature *(44,45)*, but successful agents that target other sites are emerging, in particular therapeutic antibodies that target the extracellular domain of TK receptors. In the case of ATP site, PTK-catalyzed transfer of phosphate is bisubstrate, with ATP serving as the phosphoryl donor to Tyr-containing peptide acceptors. Binding affinity is dictated by pTyr residues, which present a defining phosphoryl group that provides an array of geometrically spaced heteroatoms bearing a net (−2) charge at physiological pH, as well as amino acid residues surrounding the pTyr residue. The availability of X-ray crystal structures of SH2 domains, PTB domains, and PTPs in complex with pTyr-bearing peptides has clarified the roles of these various structural features in binding. Mutations in these regions can affect their conformation and hence their affinity to TK inhibitors, or the pathway that is either regulated by or that regulates the kinase. For example, Bardelli et al. *(46)* demonstrate the prevalence of non-synonomous mutations in several TKs involved in colorectal cancer. The missense T315I mutation identified in the Bcr-Abl from relapsed patients treated with imatinib mesylate alters both the 3-D structure of the protein-binding site and other protein–drug interactions, thus decreasing the protein sensitivity to the drug *(47)* . Mutations in the Bcr oligomerization domain can change orientation specificity and lead to the formation of higher order oligomers *(48)*. On the other hand, constitutive activity of Asp816Val mutant Kit tyrosine kinase can lead to resistance to Gleevec *(49)*.

Recalling clinical observation with Gleevec, overall mutations have been identified in more than 20 different positions within the Abl kinase domain and confer various levels of Gleevec resistance (reviewed in *50,51*), and mutations in the kinase domain of Bcr-Abl that impair Gleevec binding occur in over 50% of patients with acquired resistance to this drug *(52–55)*. Therapeutic outcome is not solely dependent on the targeted kinase, but also the context in which the mutation occurs. For instance, blast phase CML has far more heterogeneous karyotypes compared to chronic phase CML. This suggests that although p210Bcr-Abl is critical for the inception of the leukemogenic process, further mutations are involved in blast phase CML. Therefore, the success of clinical pharmacoproteomics in exploiting the impact of mutations on protein conformation, phosphorylation, and downstream protein circuits hinges on the development of novel methods for surveying the TK landscape in greater detail.

4. PROTEOMIC PROFILING OF TYROSINE KINASES

Genotyping patients for specific mutations is currently the only feasible technology for large-scale analysis of PTK. However, it is unlikely that genotyping alone is sufficient in fully predicting resistance or sensitivity. In particular, the existence of complex and unexpected crosstalk between signaling pathways may influence TK activity *(56–59)*, suggesting that secondary PTKs could be involved. In fact, resistance or sensitivity due to specific mutations of the receptors would be expected to be secondary from downstream components that are either regulators or effectors of phosphorylated and dephosphorylated circuitry. Here, phosphoproteomic analysis provides an additional tool in mutational profiling. Several approaches have been reported for phosphoprotein studies such as stable isotope labeling *(60)*, stable isotope labeling by amino acids in culture *(61,62)*, the use of antiphosphotyrosine antibodies to enrich for tyrosine-phosphorylated proteins *(63)*, high resolution 2D-MS *(64)*, high-throughput antibody/protein chip, aptamer technology *(65)*, and peptide arrays *(66)*. Details of some examples of data acquired using these techniques are given in Table 1.

With the advent of proteomic arrays, alternative technologies are emerging with the potential of developing into routine applications. In the majority of cases, allele-specific PCR has been used. However, in cases where protein-level information is required, phosphoproteomic data are needed. Clinical phosphoproteomics then becomes highly relevant in the emerging era of patient-tailored molecular medicine and could aid in identifying patients susceptible to drug treatment or to relapse.

4.1. Highlights in Phosphoproteomic Analysis of Tyrosine Kinase Activity

In cases where protein-level information is required, phosphoproteomic becomes essential. Advances in proteomic technology have been used to mine the phosphoproteome for cancer biomarkers. For example, Zhang et al. *(67)* used proteomic techniques to study the expression of proteins in six frozen tissues from breast cancer patients. Three of the six patients were positive for ErbB-2, which suggests that defects in tyrosine kinase activity were involved. Using laser-capture microdisseciton, two-dimensional gel electrophoresis (2D-GE) and matrix-assisted laser desorption and ionization coupled with time-of-flight mass spectrometry (MALDI-TOF-MS), the authors determined that cyclekeratin-19 was upregulated in ErbB2-positive patients.

Table 1
Current Proteomics Technologies for Tyrosine Kinase Analysis. Techniques for Isolation of Phosphoproteins & Phosphopeptides

Format	Target	Summary of results
Antibodies + LC-MS/MS	Phosphotyrosine-containing peptides	185 phosphotyrosine sites in c-Src expressing NIH/3T3 cells, including Src, FAK, p130Cas, cortactin, Annexin A2, and STAM–interactin protein Hrs (74)
2D-DIGE-MALDI-ToF	Proteins related to TrkA/TrkB	22 and 9 differentially expressed proteins found in SH-SY5Y neuroblastoma cells expressing TrkA or TrkB, respectively. Proteins were mainly related to maintenance of cellular structure (79)
LCM-LC-MS/MS	Non-targeted proteomics	Proof on concept study demonstrates that small ($<10^4$) cells can be analyzed. HER-2/3/4, FAK, JAK-1/3, and YES were identified (80)
Immobilised inhibitor-LC-MS/MS	Pryido [2,3-d]pyrimidin-binding (Gleevec class of compounds) proteins in SKBR-3 cells	HER-2/3/4, YES, FAK, and JAK-1/3 (81)
FLAG-IP-LC-MS/MS	Proteins interacting with MLK3	Hsp90/p50^{cdc37} interacts with kinase domain of MLK3 and regulates JNK signaling at the MAPKK level (82)

However, the use of 2D gel electrophoresis and MALDI-TOF mass spectrometry is not amenable to detailed phosphoproteomic analysis. Therefore, 2D-GE-based analysis is of limited use for phosphoproteomic analysis although it has been used effectively as a preliminary screen when supplemented with antibody-based techniques by using antibodies for both phosphorylated and non-phosphorylated forms of TKs *(68)*.

A more promising approach is the use of automated gel-free approaches to proteomic analysis. For example, the role of lasp-1 (lin, actin, and SH3-binding protein) was investigated by using multidimensional protein identification technology (MudPIT) analysis of pseudopodia in NIH 3T3 cells *(69)*. MudPIT analysis, which is an automated proteomic technique that greatly increases sensitivity and speed compared to 2D-GE analysis, allowed identification of several cytoskeletal and signal proteins in pseudopodia and allowed the determination that Lasp-1 was phosphorylated by Abl, albeit with the aid of traditional immunological techniques.

Increasingly, a more complete picture of phosphotyrosine circuitry is emerging as several powerful phosphoproteomic techniques are being coupled and applied to the study of both cell culture and patient-derived samples. The combination of two powerful techniques, tandem affinity purification (TAP) and MudPIT, has been successfully used to study the proteomic network of the protein 14-3-3σ *(70)*. Expression of this regulatory protein has been found to both increase *(71,72)* and decrease in various cancers *(73)*, which supports the notion that a more complete analysis of the phosphoproteomic status of each component in the protein circuits is required to fully understand cancer progression. The simple determination of protein upregulation or downregulation, while informative, is not sufficient to determine what downstream effects are occurring. Nevertheless, the TAP-MudPIT approach was able to identify interactions between 14-3-3σ and several kinases involved in mitogenic signaling (A-RAF1, B-RAF, c-RAF, MEKK2b, TNK1, MARK1, and MARK2). Another powerful approach is the use of immunoaffinity enrichment of phosphotyrosine-containing peptides and LC-MS/MS for proteomic profiling. When applied to Jurkat cells, this technique allowed for the identification of 194 phosphotyrosine sites, including 13 TK phosphorylation sites in p21cdc42Hs, anaplastic lymphoma kinase, and Janus kinase 3 *(74)*. Several TK substrates, specifically 10 adaptor proteins, were also determined using this technique.

4.2. Protein Microarrays

Once PTKs and other components of phosphotyrosine circuitry are identified using techniques such as MudPIT, their validation requires an increased level of throughput, sensitivity, and quantitation accuracy. This goal may potentially be achieved through the use of protein microarrays, a rapidly developing technology spawned from the success of cDNA microarrays (reviewed in ref. *75*). Unfortunately, the difficulties in producing functional proteins and the differences in experimental conditions required for protein binding currently preclude the development of whole-proteome microarrays. However, when performed in a targeted manner, protein microarrays can be a very powerful technique for probing protein expression. For example, Hudelist et al. *(76)* used a commercial antibody array consisting of 378 monoclonal antibodies to probe protein expression in breast cancer. Several kinases, including MAPKK 7, MAPKK 2, MAPKK 3, and ErbB2, were found to be upregulated in the malignant tissue relative to normal tissue, both of which were obtained from during a mastectomy procedure for a patient suffering from early breast cancer. Although the upregulation of these proteins is an important contribution to the proteomic events involved in breast cancer, the phosphorylation status of the proteins studied is conspicuously absent. For example, the use of the protein microarray did not reveal upregulation of bcl-2, despite the fact that its phosphorylation is known to be involved in taxol-induced apoptosis *(77)*. Therefore, the potential of antibody arrays will require the ability to detect not only the amount of protein present but also the changes in phosphorylation status.

Our intention in describing some recent advances in proteomic and phosphoproteomic analysis is to introduce the reader to the potential of these techniques and to elucidate their role in studying tyrosine kinase circuitry. Although these powerful techniques have provided many novel findings and improved our understanding of TKs, they are still techniques under development. In particular, no method for determination of the phosphorylation status of each component in TK circuitry currently exists. However, great strides toward this goal have been made and, on a targeted level, several intriguing observations have been made using a combination of novel proteomic technologies and more traditional techniques such as anti-pY antibodies.

4.3. Clinical Phosphoproteomics of PTKs

Insights gained from the clinical experiences with tyrosine kinase-targeted agents have helped to forsee the utility, difficulties, and challenges that might be expected to arise when using targeted therapies. Several seminal studies have established the connection between some types of mutations and the susceptibility or resistance to targeted therapies. Clinical proteomics is becoming highly relevant to the emerging era of patient-tailored molecular medicine and could aid in identifying protein features that better predict patient's response and relapse to targeted agents. An interesting example described above is the use of high-throughput sequencing technologies combined with bioinformatics to systematically analyze the tyrosine kinome in colorectal cancer *(46)*. Furthermore, proteomic techniques could be critical in defining the initial phosphoproteomic state of tumors in patients enrolled in clinical trials with kinase inhibitors. Additionally, for current TK-targeted therapies, phosphoproteomic status may be important in establishing those patients that are likely to respond. Detecting the phosphorylation status of a particular protein has been shown to be clinically relevant; the phosphorylation status of survivin, a substrate of $p34^{cdc2}$ kinase, has been shown to affect the efficacy of chemotherapy *(78)*. The availability of a wide range of specific phosphoantibodies may help expand these observations into clinical applications.

Yet, immediate challenges that remain to be solved are how to deal with tissue sampling, tissue heterogeneity, and how to standarize proteomic technology for routine use.

ACKNOWLEDGMENTS

Work from the laboratories of the authors has been supported by the Canadian Breast Cancer Research Alliance, the Canadian Institutes for Health Research, the US Army Medical Acquisition Activity, the Cancer Research Society Inc, and the National Research Council of Canada. Alaoui-Jamali is an FRSQ Scholar and Dundi and Lyon Sachs distinguished scientist.

REFERENCES

1. Robinson DR, Wu YM, Lin SF. The protein tyrosine kinase family of the human genome. Oncogene 2000;19:5548–5557.
2. Dancey J, Sausville EA. Issues and progress with protein kinase inhibitors for cancer treatment. Nat Rev Drug Discov 2003;2:296–313.
3. Hanks SK, Quinn AM. Protein kinase catalytic domain sequence database: identification of conserved features of primary structure and classification of family members. Methods Enzymol 1991;200:38–62.

4. LaRochelle WJ, Sakaguchi K, Atabey N, Cheon HG, Takagi Y, Kinaia T, Day RM, Miki T, Burgess WH, Bottaro DP. Heparan sulfate proteoglycan modulates keratinocyte growth factor signaling through interaction with both ligand and receptor. Biochemistry 1999;38:1765–1771.
5. Yarden Y, Sliwkowski MX. Untangling the ErbB signalling network. Nat Rev Mol Cell Biol 2001;2:127–137.
6. Hynes NE, Lane HA. ErbB receptors and cancer: the complexity of targeted inhibitors. Nat Rev Cancer 2005;5:341–354.
7. Plotnikov AN, Schlessinger J, Hubbard SR, Mohammadi M. Structural basis for FGF receptor dimerization and activation. Cell 1999;98:641–650.
8. Plotnikov AN, Hubbard SR, Schlessinger J, Mohammadi M. Crystal structures of two FGF-FGFR complexes reveal the determinants of ligand-receptor specificity. Cell 2000;101:413–424.
9. Stauber DJ, DiGabriele AD, Hendrickson WA. Structural interactions of fibroblast growth factor receptor with its ligands. Proc Natl Acad Sci USA 2000;97:49–54.
10. Murali R, Brennan PJ, Kieber-Emmons T, Greene MI. Structural analysis of p185c-neu and epidermal growth factor receptor tyrosine kinases: oligomerization of kinase domains. Proc Natl Acad Sci USA 1996;93:6252–6257.
11. Sicheri F, Yang DS. Structure determination of a lone alpha-helical antifreeze protein from winter flounder. Acta Crystallogr D Biol Crystallogr 1996;52: 486–498.
12. Hubbard SR. Structural analysis of receptor tyrosine kinases. Prog Biophys Mol Biol 1999;71:343–358.
13. Weiss A, Schlessinger J. Switching signals on or off by receptor dimerization. Cell 1998;94:277–280.
14. Ostman A, Bohmer FD. Regulation of receptor tyrosine kinase signaling by protein tyrosine phosphatases. Trends Cell Biol 2001;11:258–266.
15. Blume-Jensen P, Hunter T. Oncogenic kinase signalling. Nature 2001;411: 355–365.
16. Wang Z, Shen D, Parsons DW, Bardelli A, Sager J, Szabo S, Ptak J, Silliman N, Peters BA, van der Heijden MS, Parmigiani G, Yan H, Wang TL, Riggins G, Powell SM, Willson JK, Markowitz S, Kinzler KW, Vogelstein B, Velculescu VE, Mutational analysis of the tyrosine phosphatome in colorectal cancers. Science 2004;304:1164–1166.
17. Ghiglione C, Carraway KL III, Amundadottir LT, Boswell RE, Perrimon N, Duffy JB. The transmembrane molecule kekkon 1 acts in a feedback loop to negatively regulate the activity of the Drosophila EGF receptor during oogenesis. Cell 1999;96:847–856.
18. Gaestel M. MAPKAP kinases - MKs - two's company, three's a crowd. Nat Rev Mol Cell Biol 2006;7(2):120–130.
19. Papin JA, Hunter T, Palsson BO, Subramaniam S. Reconstruction of cellular signalling networks and analysis of their properties. Nat Rev Mol Cell Biol 2005;6:99–111.
20. York JD, Hunter T. Signal transduction. Unexpected mediators of protein phosphorylation. Science 2004;306:2053–2055.
21. Schulze WX, Deng L, Mann M. Phosphotyrosine interactome of the ErbB-receptor kinase family. Mol Syst Biol 2005;25:E1–E12.
22. Stephens P, Edkins S, Davies H, Greenman C, Cox C, Hunter C, Bignell G, Teague J, Smith R, Stevens C, O'Meara S, Parker A, Tarpey P, Avis T, Barthorpe A, Brackenbury L, Buck G, Butler A, Clements J, Cole J, Dicks E,

Edwards K, Forbes S, Gorton M, Gray K, Halliday K, Harrison R, Hills K, Hinton J, Jones D, Kosmidou V, Laman R, Lugg R, Menzies A, Perry J, Petty R, Raine K, Shepherd R, Small A, Solomon H, Stephens Y, Tofts C, Varian J, Webb A, West S, Widaa S, Yates A, Brasseur F, Cooper CS, Flanagan AM, Green A, Knowles M, Leung SY, Looijenga LH, Malkowicz B, Pierotti MA, Teh B, Yuen ST, Nicholson AG, Lakhani S, Easton DF, Weber BL, Stratton MR, Futreal PA, Wooster R. A screen of the complete protein kinase gene family identifies diverse patterns of somatic mutations in human breast cancer. Nat Genet 2005;37:590–592.

23. Shet AS, Jahagirdar BN, Verfaillie CM. Chronic myelogenous leukemia: mechanisms underlying disease progression. Leukemia 2002;16:1402–1411.

24. O'Brien SG, Guilhot F, Larson RA, et al. Imatinib compared with interferon and low-dose cytarabine for newly diagnosed chronic-phase chronic myeloid leukemia. N Engl J Med 2003;348:994–1004.

25. Goldman JM, Melo JV. Chronic myeloid leukemia–advances in biology and new approaches to treatment. N Engl J Med 2003;349:1451–1464.

26. Gorre ME, Mohammed M, Ellwood K, Hsu N, Paquette R, Rao PN, Sawyers CL. Clinical resistance to STI-571 cancer therapy caused by BCR-ABL gene mutation or amplification. Science 2001;293:876–880.

27. Sawyers CL. Cancer treatment in the STI571 era: what will change. J Clin Oncol 2001;19:13–16.

28. Mauro MJ, O'Dwyer M, Heinrich MC, Druker BJ. STI571: a paradigm of new agents for cancer therapeutics. J Clin Oncol 2002;20:325–334.

29. Heinrich MC, Blanke CD, Druker BJ, Corless CL. Inhibition of KIT tyrosine kinase activity: a novel molecular approach to the treatment of KIT-positive malignancies. J Clin Oncol 2002;20:1692–1703.

30. Roche-Lestienne C, Preudhomme C. Mutations in the ABL kinase domain pre-exist the onset of imatinib treatment. Semin Hematol 2003;40:80–82.

31. Roche-Lestienne C, Soenen-Cornu V, Grardel-Duflos N, Lai JL, Philippe N, Facon T, Fenaux P, Preudhomme C. Several types of mutations of the Abl gene can be found in chronic myeloid leukemia patients resistant to STI571, and they can pre-exist to the onset of treatment. Blood 2002;100:1014–1018.

32. Debiec-Rychter M, Dumez H, Judson I, Wasag B, Verweij J, Brown M, Dimitrijevic S, Sciot R, Stul M, Vranck H, Scurr M, Hagemeijer A, van Glabbeke M, van Oosterom AT; EORTC Soft Tissue and Bone Sarcoma Group. Use of c-KIT/PDGFRA mutational analysis to predict the clinical response to imatinib in patients with advanced gastrointestinal stromal tumours entered on phase I and II studies of the EORTC Soft Tissue and Bone Sarcoma Group. Eur J Cancer 2004;40:689–695.

33. Heinrich MC, Corless CL, Demetri GD, Blanke CD, von Mehren M, Joensuu H, McGreevey LS, Chen CJ, Van den Abbeele AD, Druker BJ, Kiese B, Eisenberg B, Roberts PJ, Singer S, Fletcher CD, Silberman S, Dimitrijevic S, Fletcher JA. Kinase mutations and imatinib response in patients with metastatic gastrointestinal stromal tumor. J Clin Oncol 2003;21:4342–4349.

34. Slamon DJ, Clark GM, Wong SG, Levin WJ, Ullrich A, McGuire WL. Human breast cancer: correlation of relapse and survival with amplification of the HER-2/neu oncogene. Science 1987;235:177–182.

35. Cobleigh MA, Vogel CL, Tripathy D, et al. Multinational study of the efficacy and safety of humanized anti-HER2 monoclonal antibody in women who have HER2-

overexpressing metastatic breast cancer that has progressed after chemotherapy for metastatic disease. J Clin Oncol 1999;17:2639–2648.
36. Slamon D, Pegram M. Rationale for trastuzumab (Herceptin) in adjuvant breast cancer trials. Semin Oncol 2001;28:13–19.
37. Allen LF, Eiseman IA, Fry DW, Lenehan PF. CI-1033, an irreversible pan-erbB receptor inhibitor and its potential application for the treatment of breast cancer. Semin Oncol 2003;30:65–78.
38. Malden LT, Novak U, Kaye AH, Burgess AW. Selective amplification of the cytoplasmic domain of the epidermal growth factor receptor gene in glioblastoma multiforme. Cancer Res 1988;48:2711–2714.
39. Paez JG, Janne PA, Lee JC, Tracy S, Greulich H, Gabriel S, Herman P, Kaye FJ, Lindeman N, Boggon TJ, Naoki K, Sasaki H, Fujii Y, Eck MJ, Sellers WR, Johnson BE, Meyerson M. EGFR mutations in lung cancer: correlation with clinical response to gefitinib therapy. Science 2004;304:1497–1500.
40. Lynch TJ, Bell DW, Sordella R, et al. Activating mutations in the epidermal growth receptor underlying responsiveness of non-small-cell lung cancer to Gefitinib. N Engl J Med 2004;350:2129–2139.
41. Kobayashi S, Boggon TJ, Dayaram T, Janne PA, Kocher O, Meyerson M, Johnson BE, Eck MJ, Tenen DG, Halmos B. EGFR mutation and resistance of non-small-cell lung cancer to gefitinib. N Engl J Med 2005;352:786–792.
42. Stephens P, Edkins S, Davies H, et al. A screen of the complete protein kinase gene family identifies diverse patterns of somatic mutations in human breast cancer. Nat Genet 2005;37:590–592.
43. Stephens P, Hunter C, Bignell G, et al. Lung cancer: intragenic ERBB2 kinase mutations in tumours. Nature 2004;431:525–526.
44. Morin MJ. From oncogene to drug: development of small molecule tyrosine kinase inhibitors as anti-tumor and anti-angiogenic agents. Oncogene 2000;19:6574–6583.
45. Traxler P, Bold G, Buchdunger E, Caravatti G, Furet P, Manley P, O'Reilly T, Wood J, Zimmermann J. Tyrosine kinase inhibitors: from rational design to clinical trials. Med Res Rev 2001;21:499–512.
46. Bardelli A, Parsons DW, Silliman N, Ptak J, Szabo S, Saha S, Markowitz S, Willson JK, Parmigiani G, Kinzler KW, Vogelstein B, Velculescu VE. Mutational analysis of the tyrosine kinome in colorectal cancers. Science 2003;300:949.
47. Pricl S, Fermeglia M, Ferrone M, Tamborini E. T315I-mutated Bcr-Abl in chronic myeloid leukemia and imatinib: insights from a computational study. Mol Cancer Ther 2005;4:1167–1174.
48. Taylor CM, Keating AE. Orientation and oligomerization specificity of the Bcr coiled-coil oligomerization domain. Biochemistry 2005;44:16246–16256.
49. Foster R, Griffith R, Ferrao P, Ashman L. Molecular basis of the constitutive activity and STI571 resistance of Asp816Val mutant KIT receptor tyrosine kinase. J Mol Graph Model 2004;23:139–152.
50. Deininger M, Buchdunger E, Druker BJ. The development of imatinib as a therapeutic agent for chronic myeloid leukemia. Blood 2005;105:2640–2653.
51. Deininger M. Resistance to imatinib: mechanisms and management. J Natl Compr Canc Netw 2005;3:757–768.
52. Gorre ME, Mohammed M, Ellwood K, Hsu N, Paquette R, Rao PN, Sawyers CL. Clinical resistance to STI-571 cancer therapy caused by BCR-ABL gene mutation or amplification. Science 2001;293:876–880.
53. Shah NP, Nicoll JM, Nagar B, Gorre ME, Paquette RL, Kuriyan J, Sawyers CL. Multiple BCR-ABL kinase domain mutations confer polyclonal resistance to the

tyrosine kinase inhibitor imatinib (STI571) in chronic phase and blast crisis chronic myeloid leukemia. Cancer Cell 2002;2:117–125.
54. von Bubnoff N, Schneller F, Peschel C, Duyster J. BCR-ABL gene mutations in relation to clinical resistance of Philadelphia-chromosome-positive leukaemia to STI571: a prospective study. Lancet 2002;359:487–491.
55. Al-Ali HK, Heinrich MC, Lange T, Krahl R, Mueller M, Muller C, Niederwieser D, Druker BJ, Deininger MW. High incidence of BCR-ABL kinase domain mutations and absence of mutations of the PDGFR and KIT activation loops in CML patients with secondary resistance to imatinib. Hematol J 2004;5:55–60.
56. Nagata Y, Lan KH, Zhou X, Tan M, Esteva FJ, Sahin AA, Klos KS, Li P, Monia BP, Nguyen NT, Hortobagyi GN, Hung MC, Yu D. PTEN activation contributes to tumor inhibition by trastuzumab, and loss of PTEN predicts trastuzumab resistance in patients. Cancer Cell 2004;6:117–127.
57. Yamauchi T, Ueki K, Tobe K, Tamemoto H, Sekine N, Wada M, Honjo M, Takahashi M, Takahashi T, Hirai H, Tushima T, Akanuma Y, Fujita T, Komuro I, Yazaki Y, Kadowaki T. Tyrosine phosphorylation of the EGF receptor by the kinase Jak2 is induced by growth hormone. Nature 1997;390:91–96.
58. Andreev J, Galisteo ML, Kranenburg O, Logan SK, Chiu ES, Okigaki M, Cary LA, Moolenaar WH, Schlessinger J. Src and Pyk2 mediate G-protein-coupled receptor activation of epidermal growth factor receptor (EGFR) but are not required for coupling to the mitogen-activated protein (MAP) kinase signaling cascade. J Biol Chem 2001;276:20130–20135.
59. Biscardi JS, Maa MC, Tice DA, Cox ME, Leu TH, Parsons SJ. c-Src-mediated phosphorylation of the epidermal growth factor receptor on Tyr845 and Tyr1101 is associated with modulation of receptor function. J Biol Chem 1999;274: 8335–8343.
60. Zhang Y, Wolf-Yadlin A, Ross PL, Pappin DJ, Rush J, Lauffenburger DA, White FM. Time-resolved mass spectrometry of tyrosine phosphorylation sites in the epidermal growth factor receptor signaling network reveals dynamic modules. Mol Cell Proteomics 2005;4:1240–1250.
61. Ong S-E, Blagoev B, Kratchmarova I, Kristensen DB, Steen H, Pandey A, Mann M. Stable isotope labeling by amino acids in cell culture, SILAC, as a simple and accurate approach to expression proteomics. Mol Cell Proteomics 2002;1:376–386.
62. Hinsby AM, Olsen JV, Mann M. Tyrosine phosphoproteomics of fibroblast growth factor signaling: a role for insulin receptor substrate-4. J Biol Chem 2004;279:46438–46447.
63. Cutillas PR, Geering B, Waterfield MD, Vanhaesebroeck B. Quantification of gel-separated proteins and their phosphorylation sites by LC-MS using unlabeled internal standards: analysis of phosphoprotein dynamics in a B cell lymphoma cell line. Mol Cell Proteomics 2005;4:1038–1051.
64. Lim YP, Diong LS, Qi R, Druker BJ, Epstein RJ. Phosphoproteomic fingerprinting of epidermal growth factor signaling and anticancer drug action in human tumor cells. Mol Cancer Ther 2003;2:1369–1377.
65. Crawford M, Woodman R, Ferrigno PK. Peptide aptamers: tools for biology and drug discovery. Brief Funct Genomic Proteomic 2003;2:72–79.
66. Manke IA, Lowery DM, Nguyen A, Yaffe MB. BRCT repeats as phosphopeptide-binding modules involved in protein targeting. Science 2003;302:636–639.
67. Zhang D-H, Tai LK, Wong LL, Sethi SK, Koay ESC. Proteomics of breast cancer: enhanced expression of cytokeratin19 in human epidermal growth factor receptor type 2 positive breast tumors. Proteomics 2005;5:1797–1805.

68. Imamura T, Kanai F, Kawakami T, Amarsanaa J, Ijichi H, Hoshida Y, Tanaka Y, Ikenoue T, Tateishi K, Kawabe T, Arakawa Y, Miyagishi M, Taira K, Yokosuka O, Omata M. Proteomic analysis of the TGF-beta signaling pathway in pancreatic carcinoma cells using stable RNA interference to silence Smad4 expression. Biochem Biophys Res Commun 2004;318:289–296.
69. Lin YH, Park ZY, Lin D, Brahmbhatt AA, Rio MC, Yates JR III, Klemke RL. Regulation of cell migration and survival by focal adhesion targeting of Lasp-1. J Cell Biol 2004;165:421–432.
70. Benzinger A, Muster N, Koch HB, Yates JR III, Hermeking H. Targeted proteomic analysis of 14-3-3 sigma, a p53 effector commonly silenced in cancer. Mol Cell Proteomics 2005;4:785–795.
71. Liu Y, Liu H, Han B, Zhang JT. Identification of 14-3-3 sigma as a contributor to drug resistance in human breast cancer cells using functional proteomic analysis. Cancer Res 2006;66:3248–3255.
72. Han B, Xie H, Chen Q, Zhang JT. Sensitizing hormone-refractory prostate cancer cells to drug treatment by targeting 14-3-3sigma. Mol Cancer Ther 2006;5: 903–912.
73. Hermeking H. The 14-3-3 cancer connection. Nat Rev Cancer 2003;3:931–943.
74. Rush J, Moritz A, Lee KA, Guo A, Goss VL, Spek EJ, Zhang H, Zha X-M, Polakiewicz RD, Comb MJ. Immunoaffinity profiling of tyrosine phosphorylation in cancer cells. Nat Biotechnol 2005;23:94–101.
75. Kingsmore SF. Multiplexed protein measurement: technologies and applications of protein and antibody arrays. Nat Rev Drug Discov 2006;5:310–320.
76. Hudelist G, Pacher-Zavisin M, Singer CF, Holper T, Kubista E, Schreiber M, Manavi M, Bilban M, Czerwenka K. Use of high-throughput protein array for profiling of differentially expressed proteins in normal and malignant breast tissue. Breast Cancer Res Treat 2004;86:281–291.
77. Huang Y, Sheikh MS, Fornace AJ Jr, Holbrook NJ. Serine protease inhibitor TPCK prevents Taxol-induced cell death and blocks c-Raf-1 and Bcl-2 phosphorylation in human breast carcinoma cells. Oncogene 1999;18:3431–3439.
78. O'Connor DS, Wall NR, Porter AC, Altieri DC. A p34(cdc2) survival checkpoint in cancer. Cancer Cell 2002;2:43–54.
79. Sitek B, Apostolov O, Sitek B, Apostolov O, Stühler K, Pfeiffer K, Meyer HE, Eggert A, Schramm A. Identification of dynamic proteome changes upon ligandactivation of Trk-receptors using two-dimensional fluorescence difference gel electrophoresisand mass spectrometry. Mol Cell Proteomics 2005;4(3):291–299.
80. Wu S-L, Hancock WS, Goodrich GG, Kunitake ST. An approach to the proteomic analysis of a breast cancer cell line (SKBR-3). Proteomics 2003;3:1037–1046.
81. Wissing J, Godl K, Brehmer D, Blencke S, Weber M, Habenberger P, Stein-Gerlach M, Missio A, Cotton M, Müller S, Daube H. Chemical proteomic analysis reveals alternative modes of action for pyrido[2,3-d]pyrimidine kinase inhibitors. Mol Cell Proteomics 2004;3(12):1181–1193.
82. Zhang H, Wu W, Du Y, Santos SJ, Conrad SE, Watson JT, Grammatikakis N, Gallo KA. Hsp90/p50cdc37 is required for mixed-lineage kinase(MLK) 3 signaling. J Biol Chem 2004;279(19):19457–19463.

III Tumor Proteomics

4 Oncoproteomics for Personalized Management of Cancer

K. K. Jain

CONTENTS

1. INTRODUCTION
2. ROLE OF PROTEOMICS IN DISCOVERY OF BIOMARKERS OF CANCER
3. APPLICATION OF PROTEOMICS IN ONCOLOGY
4. ROLE OF PROTEOMICS IN CLINICAL TRIALS OF PERSONALIZED ANTICANCER DRUGS
5. ROLE OF PROTEOMICS IN MANAGEMENT OF VARIOUS CANCERS
6. CONCLUSIONS

SUMMARY

This chapter describes the role of proteomics in developing personalized management of cancer, which is defined as choice of the best treatment for an individual. Numerous proteomic technologies are available, and selected ones relevant to development of personalized therapy are described. Cancer biomarkers discovered by proteomics can be used for diagnosis as well as drug targets, thus facilitating the integration of diagnostics and therapeutics. Proteomics fulfills many of the requirements for personalized therapy of cancer.

From: *Cancer Drug Discovery and Development
Cancer Proteomics: From Bench to Bedside*
Edited by: S. S. Daoud © Humana Press Inc., Totowa, NJ

Key Words: Anticancer drugs; biomarkers; cancer; protein microarrays; oncoproteomics; personalized medicine; pharmacogenomics; pharmacoproteomics; protein profiling; proteomics

1. INTRODUCTION

Oncoproteomics is the term used for application of proteomic technologies in oncology, and this chapter will discuss the role it plays in the development of personalized medicine, which simply means selection of therapy best suited for an individual patient *(1)*. Pharmacogenomics (application of genomics to drug discovery and development), pharmacogenetics (influence of genetic factors on action of drugs), and pharmacoproteomics (application of proteomics to drug discovery and development) play an important role in this process *(2)*. However, metabolomics (study of metabolites) and environmental factors also need to be taken into consideration. Management of cancer has been unsatisfactory in the past, but an understanding of the molecular, genetic, genomic, and proteomic aspects of cancer is accelerating progress in cancer therapy. A major improvement in cancer therapy is the introduction of the concept of personalization. Personalized medicine requires a better understanding of the molecular biology of the disease, improved diagnosis for early detection as well as monitoring the course of the disease, and rational therapeutics, which may be combined with appropriate diagnostics. The role of proteomics in the development of personalized therapy for cancer is shown in Fig. 1.

This chapter will describe the use of proteomics for developing personalized therapies for cancer. Proteomic technologies have been described in detail elsewhere *(3)*. Selected proteomic technologies will facilitate the development of personalized medicine as follows:

- a better understanding of cancer molecular pathology of cancer;
- improving the molecular diagnosis of cancer;
- improved classification of cancer;
- cancer biomarkers discovered by proteomics can be used for diagnosis as well as drug targets, thus facilitating the integration of diagnostics and therapeutics;
- toxicoproteomics, by identifying toxic effects of anticancer drugs at an early stage, will help in the development of safer therapies for cancer.

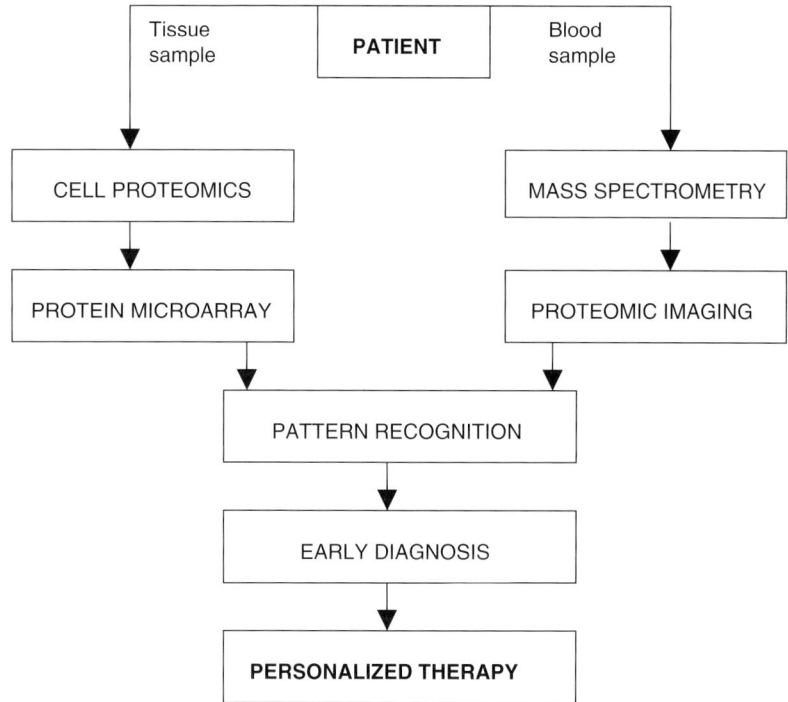

Fig. 1. Role of proteomics in development of personalized therapy for cancer.

2. ROLE OF PROTEOMICS IN DISCOVERY OF BIOMARKERS OF CANCER

Any measurable specific molecular alteration of a cancer cell on either DNA, RNA, or protein level can be referred to as a biomarker, for example, oncoproteins are biomarkers for cancer. In recent years, the discovery of cancer biomarkers has become a major focus of cancer research. The widespread use of prostate-specific antigen (PSA) in prostate cancer screening has motivated researchers to identify suitable markers for screening different types of cancer. Biomarkers are also useful for diagnosis, monitoring cancer progression, predicting recurrence, and assessing efficacy of treatment. The advent of targeted therapies such as imatinib (Novartis' Gleevec), trastuzumab (Roche's Herceptin), and rituximab (Roche's Mabthera), where a causal relationship has been established between drug target and therapy, drives the need for biomarkers for selecting the patients for a given therapy as well as for predicting drug resistance. Biomarkers are thus important for the development of personalized medicine. Proteins may be actively secreted or released by the tumor cells as a result of necrosis or apoptosis and released into the circulation. They change the protein profile. The difference in signal intensities may be detected by comparison with sera from normal individuals. Examples of cancer assays based on proteins and enzymes are (1) P53 sequencing and functional assays; (2) measurement of telomerase activity; (3) prognostic assay based on survivin (inhibitor of apoptosis protein). Proteomic technologies used for the discovery of cancer biomarkers are listed in Table 1.

2.1. Antibody Microarrays

The use of antibody microarrays continues to grow rapidly due to the recent advances in proteomics and automation and the opportunity this combination creates for high-throughput multiplexed analysis of protein biomarkers. However, a primary limitation of this technology is the lack of polymerase chain reaction-like amplification methods for proteins. Therefore to realize the full potential of array-based protein biomarker screening, it is necessary to construct assays that can detect and quantify protein biomarkers with very high sensitivity, in the femtomolar range and from limited sample quantities. Ultramicroarrays (BioForce Nanosciences Inc) combine the advantages of microarraying including multiplexing capabilities, higher throughput and cost savings, with the ability to screen very small sample volumes *(4)*. Antibody ultramicroarrays for the detection of interleukin-6 and PSA, a widely

Table 1
Proteomic Technologies Used for the Discovery of Cancer Biomarkers

Proteomic technologies
Antibody microarrays
Aptamer-based technology for protein signatures of cancer cells
CellCarta (Caprion Pharmaceuticals)
Detection of circulating nucleosomes in serum
eTag assay system for cancer biomarkers
Gel electrophoresis: 2D PAGE
Laser capture microdissection
Tissue proteomics: molecular histopathology of cancer
ProteinChip technology
Surface-enhanced laser desorption ionization time-of-flight mass spectrometry

2D Page, two-dimensional polyacrylamide gel electrophoresis.

used biomarker for prostate cancer screening, were constructed. These ultramicroarrays were found to have a high specificity and sensitivity with detection levels using purified proteins in the attomole range. Ultramicroarrays have enabled detection of PSA secreted from just four cells in 24 h. Cellular PSA could also be detected from the lysate of an average of just six cells. This strategy should enable proteomic analysis of materials that are available in very limited quantities such as those collected by laser capture microdissection (LCM).

2.2. Two-dimensional PAGE

Two-dimensional polyacrylamide gel electrophoresis (2D PAGE) followed by protein identification using mass spectrometry (MS) has been the primary technique for biomarker discovery in conventional proteomic analyses. This technique is uniquely suited for direct comparisons of protein expression and has been used to identify proteins that are differentially expressed between normal and tumor tissues in various cancers, such as liver, bladder, lung, esophageal, prostate, and breast. Advantages of 2D PAGE for discovery of biomarkers are that (1) it is a tested and reliable method; (2) it can identify markers directly; (3) it is reproducible and quantitative when combined with fluorescent dyes. Disadvantages of the use of 2D PAGE for this purpose are that (1) it requires a large amount of protein as starting material making it unreliable for detecting and identifying low-abundance proteins; (2) early stage cancers are often small and contamination from surrounding stromal tissue that is present in the specimen can confound the detection of tumor-specific markers; (3) low sensitivity. The sensitivity can be improved by LCM.

2.3. SELDI-TOF MS

Surface-enhanced laser desorption ionization time-of-flight (SELDI-TOF) MS is an important tool for the rapid identification of cancer-specific biomarkers and proteomic patterns in the proteomes of both tissues and body fluids. It is useful in high-throughput proteomic fingerprinting of cell lysates and body fluids that uses on-chip protein fractionation coupled to TOF separation. Within minutes, sub-proteomes of a complex milieu such as serum can be visualized as a proteomic fingerprint or "bar-code." SELDI technology has significant advantages over other proteomic technologies in that the amounts of input material required for analysis are miniscule compared with more traditional 2D PAGE approaches *(5)*. A number of studies have used

SELDI technology to identify single-disease-related biomarkers for several types of cancer. For example, a modified, quantitative SELDI approach has been used to show that the levels of serum prostate-specific membrane antigen are significantly higher in patients with prostate cancer than in those with benign disease.

2.4. Detection of Tumor Markers with ProteinChip Technology

The ProteinChip Biomarker System (Ciphergen) was developed for the Expression Difference Mapping™ of several hundreds of samples per day in a single uncomplicated platform with software support for the construction of multi-marker predictive models. The Interaction Discovery Mapping™ platform was next introduced for the investigation of protein-binding partners of possible importance in diagnosis and therapy. SELDI-based ProteinChip technology has been used for the detection of tumor biomarkers (6). The multimarker system has been shown to be superior over the single-marker strategy and is faster. This system has been used for the detection of biomarkers of cancers of several organs including the prostate, ovary, breast, and lungs.

SELDI-TOF-MS of platelet extracts for proteomic profiling shows increased amounts of angiogenic regulatory proteins such as vascular endothelial growth factor and endostatin in platelet but not in plasma. This is a selective sequestration process and not a simple association with the platelet surface. This novel property of platelets detects human cancers of a microscopic size undetectable by any presently available diagnostic method. This is more inclusive than a single biomarker because it can detect a wide range of tumor types and tumor sizes. Relative changes in the platelet angiogenic profile permit the tracking of a tumor throughout its development, beginning from an early in situ cancer.

2.5. Tissue Proteomics for Discovery of Cancer Biomarkers

The molecular complexity of tissue and the in vivo inaccessibility of cells within solid tumors hinder efforts to discover new diagnostic biomarkers useful for noninvasive tumor-specific molecular imaging. Scientists at the Sidney Kimmel Cancer Center (San Diego, CA), using a sub-cellular fractionation of tissue, subtractive proteomic analysis, bioinformatics, and expression profiling, have identified several biomarkers induced in solid tumors at the tissue–blood interface that are accessible to agents injected intravenously. Molecular imaging validates immunotargeting and penetration of single organs and solid tumors within an hour. These markers are expressed on vascular endothelium and its specialized transport vesicles (caveolae) in multiple human and rodent tumors including primary and metastatic lesions found in the liver, brain, breast, kidney, lung, intestine, and prostate. Mapping tissue- and disease-modulated endothelial cell surface and caveolar proteins reveals promising biomarkers for imaging and therapy of solid tumors.

2.6. eTag Assay System for Cancer Biomarkers

The eTag assay system (Monogram Biosciences) is a high performance, high-throughput system for the study of tens to hundreds of genes, proteins, and cell-based antigens across thousands of samples. The system makes it possible for researchers to adopt a systems biology approach toward studies of gene expression, protein expression, and for applications such as cell signaling and pathway activation, protein–protein interaction, and cell receptor binding. The system uses Monogram Biosciences proprietary eTag reporters to multiplex the analysis of genes and/or proteins from the same sample. Specific molecular binding events result in the release of electrophoretically distinct eTag reporters, which are then resolved to provide precise, sensitive quantification of multiple analytes directly from cell lysates.

The eTag assay system is ideally suited to analysis of complex biology and medicine, such as that seen in oncology. These unique assays can precisely measure many types of pathway biomarkers simultaneously using small samples, such as those obtained from standard tumor biopsies. Monogram Biosciences collaborating with Norris Comprehensive Cancer Center at the University of Southern California, Los Angeles to identify and characterize novel clinical biomarkers for breast cancer. These biomarkers could be used to correlate disease type and progression, resulting in improved treatment. Novel eTag assays for unique protein biomarkers such as receptor complexes and phosphorylation events will be developed to focus on profiling epidermal growth factor receptor (EGFR) family signal transduction pathways. Further research will be aimed at applying eTag assays to retrospective analysis of patient samples for validation and diagnostic development. Knowledge gained from these studies will improve the prognosis of cancer patients by better understanding of treatments that will effect disease progression in specific patients. It is expected that these innovative eTag assays will enable the development of a broad range of biomarker-based personalized medicines to improve treatment for cancer.

2.7. Detection of Circulating Nucleosomes in Serum of Cancer Patients

In the nucleus of eukaryotic cells, DNA is associated with several protein components and forms complexes known as nucleosomes. During cell death, particularly during apoptosis, endonucleases are activated that cleave the chromatin into multiple oligonucleosomes and mononucleosomes. Subsequently, these nucleosomes are packed into apoptotic bodies and are engulfed by macrophages or neighboring cells. In cases of high rates of cellular turnover and cell death, they also are released into the circulation and can be detected in serum or plasma by Cell Death Detection-ELISAplus (CDDE) from Roche Diagnostics (Mannheim, Germany). As enhanced cell death occurs under various pathologic conditions, elevated amounts of circulating nucleosomes are not specific for any benign or malignant disorder. However, the course of change in the nucleosomal levels in circulation of patients with malignant tumors during chemotherapy or radiotherapy is associated with the clinical outcome and can be useful for the therapeutic monitoring and the prediction of the therapeutic efficacy.

2.8. Detection of Tumor-Specific Marker Glycoproteins

Gels prepared by biomolecular imprinting using lectin and antibody molecules as ligands have been used for recognition for tumor-specific marker glycoproteins *(7)*. The glycoprotein-imprinted gels prepared with minute amounts of cross-linkers can dynamically recognize tumor-specific marker glycoproteins by lectin and antibody ligands and induce volume changes according to the glycoprotein concentration. The glycoprotein-imprinted gels shrink in response to a target glycoprotein but nonimprinted gels expand. The glycoprotein-responsive shrinking of the imprinted gel is caused by formation of lectin–glycoprotein–antibody complexes that act as reversible cross-linking points. Glycoprotein-imprinted gels only shrink when both lectin and antibody in the gels simultaneously recognize the saccharide and peptide chains of the target glycoprotein. As shrinking behavior of biomolecularly imprinted gels in response to glycoproteins enables the accurate detection and recognition of tumor-specific marker glycoproteins, they have many potential applications as smart devices in sensing systems and for molecular diagnostics of cancer.

3. APPLICATION OF PROTEOMICS IN ONCOLOGY

Currently, one of the most popular applications of proteomics is in the area of cancer research. Study of relevant proteins is useful for gaining an understanding of the pathology and progression of cancer. Proteomics-based approaches, which enable the quantitative investigation of both cellular protein expression levels and protein–protein interactions involved in signaling networks, promise to define the molecules controlling the processes involved in cancer. Proteomics-based profiling uniquely allows delineation of global changes in protein expression patterns resulting from transcriptional and post-transcriptional control, post-translational modifications, and shifts in proteins between different cellular compartments. Given that comprehensive expression profiles obtained using genomics and proteomics are highly complementary, a combined approach to profiling may well uncover expression patterns that could not be predicted using a single approach. Some of the proteomics technologies used in cancer that are relevant to personalized medicine will be described here.

3.1. Application of CellCarta Technology for Oncology

CellCarta™ (Caprion) is used to identify entire protein complement of organelles under normal and diseased conditions. It can deliver drug targets and functional insights unattainable by other proteomics approaches. Whereas the conventional approaches provide only incomplete protein identification showing abundant proteins only with no information on protein location, the CellCarta Cell Maps provide comprehensive protein identification including low abundance proteins with their location and orientation. Protein trafficking events are identified and there is reliable protein expression profiling. Mass intensity profiling system (MIPS) measures the intensity of peptide ions derived from a specific protein as a relative measure of protein abundance and generates comprehensive protein expression profile for each sample linked with protein identification. Advantages of MIPS include automation, comprehensive protein coverage, and amenability to "gel-free" analyses. Practical applications in drug discovery include target validation, determination of mechanism of action of leads, and toxicity profiling. CellCarta has been applied in areas of special pharmaceutical interest for tumor antigen discovery, protein biomarker discovery, pharmacoproteomics, and protein phosphorylation. Applications in oncology are the most extensive. Isolation of purified membrane proteins from resected tumors is followed by differential analysis of protein abundance. High value protein targets are identified. Examples of applications in oncology are in lung and colon cancer.

3.2. Phage Display Technology

Phage display technology utilizes combinatorial libraries of proteins expressed on phage particles that can be selected for specific binding to cancer cells (8). Such cancer-specific molecules can be used in a variety of applications, including identification of cell-specific targeting molecules, identification of cell surface biomarkers, profiling of specimens obtained from individual cancer patients, and the design of peptide-based anticancer therapeutics for personalized treatments. Peptide phage display strategies can target cell surfaces because many biomarkers important in cancer are differentially expressed molecules located on the outside of the cell membranes.

A novel method called ADEPPT (accentuation of differentially expressed proteins using phage technology) can identify proteins that are produced in different amounts in diseased tissue compared with healthy tissue (9). The study used a large "library" with thousands of strains of the bacteriophage known as M13. Each strain of M13 makes a different peptide, which binds to a specific protein. By using a large bacteriophage library, the researchers had the best possible chance of detecting all the proteins that were produced at varying levels in different tissue samples. This technique can rapidly determine differences between lung cancer tissue and normal lung tissue by measuring subtle variations in the proteins they produce. ADEPPT can pick up proteins in lung cancer that are overlooked by more conventional methods of protein profiling. This method may ultimately enable researchers to detect proteins responsible for all types of cancer and potentially assist them in finding better drug targets to treat various diseases. This information could also be used for early detection of cancers, for example, by testing for elevated levels of proteins in the blood, which could potentially improve the chance of successful treatment. Further development of the ADEPPT method is needed, but initial indications are that it may complement existing methods such as 2D gel electrophoresis by detecting less abundant proteins.

3.3. Proteomic Analysis of Cancer Cell Mitochondria

Mitochondria are essential organelles for cellular homeostasis. Mutations in mitochondrial DNA have been frequently reported in cancer cells. Proteomics-based methods have been applied not only to the analysis of protein function in the organelle but also to identify biomarkers for diagnosis and therapeutic targets of specific pathologies associated with mitochondria *(10)*. Role of proteomics in the study of mitochondrial proteome in cancer is the following:

- Identification of abnormally expressed mitochondrial proteins in cancer cells is possible by mitochondrial functional proteomics.
- Proteomics can identify new markers for early detection and risk assessment, as well as targets for therapeutic intervention.

3.4. Id Proteins as Targets for Cancer Therapy

Id (inhibitor of DNA binding) proteins represent attractive targets for cancer therapy. Their involvement in the progression of human cancer is now established, based on the analysis of tumor cell cultures, human cancer biopsies, and animal tumor models. Id proteins within invasive and progressive cancers have distinctive expression patterns that are potential specific diagnostic and/or prognostic markers *(11)*.

Multiple strategies can now be used to target the functional activities of intracellular proteins for cancer therapy. These include antisense oligonucleotides, siRNA, anti-gene or ribozyme gene transfer, and small molecules or peptides. However, each of these strategies has potential advantages and disadvantages. Even though numerous genes that are regulated by Id gene expression in cancer cells have been identified, much work must be done to link these associated genes to downstream functional activities.

3.5. Cancer Tissue Proteomics for Assessment of Therapy

Cancer tissue proteomics implies direct tissue profiling and use of imaging MALDI MS to provide a molecular assessment of numerous expressed proteins within a tissue sample. Analysis of thin tissue sections results in the visualization of 500–1000 individual protein signals in the molecular weight range from 2000 to over 200,000 *(12)*. LCM, in combination with MS, enables acquisition of protein signatures from a single cell type within a heterogeneous sample. These signals directly correlate with protein distribution within a specific region of the tissue sample. The systematic investigation of the section allows the construction of ion density maps, or specific molecular images, for virtually every signal detected in the analysis.

MALDI TOF MS can be used to generate protein spectra directly from frozen tissue sections from surgically resected cancer specimens. Profiling MALDI MS has been used to monitor alterations in protein expression associated with tumor progression and metastases. Current data suggest that MALDI MS will be superior to immunohistochemical stains and electron microscopy in identifying the site of origin for tumors currently labeled as "tumor of unknown primary." Another application in surgical pathology would be the rapid evaluation of margins of surgical excision of a tumor. Routine analysis of surgical margins by frozen section is very difficult because some cancers invade in a single cell fashion without producing a grossly identifiable mass. Sensitivity of MS enables detection of even a few tumor cells within a significantly larger portion of tissue.

The capability of MALDI MS to measure susceptibility and response to therapeutic agents in tumor and surrounding tissues is particularly useful in personalized management of cancer. The original protein profile obtained from the primary tumor can be used to influence the selection of therapeutic agents. Levels of chemotherapeutic agents can be measured directly from a tissue biopsy to assess adequacy of delivery to a particular organ site. It will also help in detecting alterations in specific molecular pathways directly modulated or indirectly affected by the anticancer agent. Finally, it could be used to monitor chemotherapy effects on the tumor.

3.6. Morphoproteomics

Morphoproteomics combines histopathology, molecular biology, and proteomics to depict the protein circuitry in diseased cells for uncovering molecular targets amenable to specific intervention, thereby facilitating personalized therapy *(13)*. Such an approach can uncover or confirm potential molecular targets that may be essential to the growth, integrity, and histogenesis of a particular tumor type and that are amenable to specific therapeutic interventions. Directions for future research should focus on points of convergence in signal transduction pathways and consider integration of morphoproteomic with genomic and pharmacoproteomic as well as protein-function microarray data.

3.7. Application of Nanobiotechnology in Oncoproteomics

Nanotechnology is the creation and utilization of materials, devices, and systems through the control of matter on the nanometer-length scale, that is, at the level of atoms, molecules, and supramolecular structures. It is the popular term for the construction and utilization of functional structures with at least one characteristic dimension measured in nanometers,? a nanometer is one billionth of a meter (10^{-9} m). Various nanobiotechnologies and their applications are described elsewhere *(14)*.

Nanoproteomics—application of nanobiotechnology to proteomics—improves on most current protocols including protein purification/display and automated identification schemes that yield unacceptably low recoveries with reduced sensitivity and speed while requiring more starting material. Low abundant proteins and proteins that can only be isolated from limited source material (e.g., biopsies of tumors) can be subjected to nanoscale protein analysis, that is, nano-capture of specific proteins and complexes, and optimization of all subsequent sample handling steps leading to mass analysis of peptide fragments. This is a focused approach, also termed targeted proteomics, and involves examination of subsets of the proteome, for example, those proteins that are either specifically modified or bind to a particular DNA sequence or exist as members of higher order complexes or any combination thereof.

An ion mobility technology—high-field asymmetric waveform ion mobility mass spectrometry (FAIMS)—has been introduced as online ion selection methods compatible with electrospray ionization. FAIMS uses ion separation to improve detection limits of peptide ions when used in conjunction with electrospray and nanoelectrospray MS. This facilitates the identification of low abundance peptide ions often present at ppm levels in complex proteolytic digests and expand the sensitivity and selectivity of nanoLC–MS analyses in global and targeted proteomics approaches. This functionality likely will play an important role in drug discovery and biomarker programs for monitoring of disease progression and drug efficacy *(15)*. It can be applied to cancer.

4. ROLE OF PROTEOMICS IN CLINICAL TRIALS OF PERSONALIZED ANTICANCER DRUGS

Therapeutics combined with companion diagnostics is considered to be a part of personalized medicine. Some of the proteomics-based diagnostics as well as therapeutics may be tested in clinical trials.

Biomarkers can aid in patient stratification (risk assessment), in treatment response identification (surrogate markers), or in differential diagnosis (identifying individuals who are likely to respond to specific drugs). To be clinically useful, a marker must favorably affect clinical outcomes such as decreased toxicity, increased overall and/or disease-free survival, or improved quality of life. Once the methods for assessment of the biomarker are established and the initial results show promise with regard to the predictive ability of a marker, it may be possible to achieve the goal of "predictive oncology."

5. ROLE OF PROTEOMICS IN MANAGEMENT OF VARIOUS CANCERS

Proteomics has an impact on both the diagnosis and treatment of cancer of various organs. Some examples are given here.

5.1. HER-2/neu Oncoprotein as Biomarkers for Breast Cancer

HER-2/neu oncoprotein has been widely studied for many years and has been shown to play a pivotal role in the development and progression of breast cancer. HER-2/neu has been shown to be an indicator of poor prognosis with patients exhibiting aggressive disease, decreased overall survival, and a higher probability of recurrence of disease. As evidenced by numerous published studies, elevated levels of HER-2/neu (also referred to as overexpression) are found in about 30% of women with breast cancer. Determination of a patient's HER-2/neu status may be valuable in identifying whether that patient has a more aggressive disease and would, thus, derive substantial benefit from more intensive or alternative therapy regimens. Elevated levels of HER-2/neu are found not only in breast cancer but also in several other tumor types including prostate, lung, pancreatic, colon, and ovarian cancers.

Some studies suggest that in certain breast cancer patients, persistently rising HER-2/neu values may be associated with aggressive cancer and poor response to therapy, while decreasing HER-2/neu levels may be indicative of effective therapy. The clinical utility of the serum test as a prognostic indicator has not yet been fully established but is under investigation.

Traditional HER-2/neu testing is generally limited to tissue from primary breast cancer and does not provide information regarding the HER-2/neu status in women with recurrent, metastatic breast cancer. The introduction of microtiter plate ELISA HER-2/neu testing (Bayer Diagnostics) using a serum sample now offers a less invasive diagnostic tool and provides a current assessment of a woman's HER-2/neu status over the course of disease.

Immunohistochemistry (IHC) analysis of HER2/neu in breast carcinoma is a useful predictor of response to therapy with trastuzumab when strongly positive. Negative immunostaining is highly concordant with a lack of gene amplification by FISH. Most weakly positive overexpressors are false positives on testing with FISH. Thus, screening of breast carcinomas with IHC and confirmation of weakly positive IHC results by FISH is an effective evolving strategy for testing HER2/neu as a predictor of response to targeted therapy.

Breast cancers show variable sensitivity to paclitaxel. Tubulin polymerization assay was used to show that low tau expression renders microtubules that are more vulnerable to paclitaxel and makes breast cancer cells hypersensitive to this drug *(16)*. Low tau expression, therefore, may be used as a marker to select patients for paclitaxel therapy. Inhibition of tau function by RNAi might be exploited as a therapeutic strategy to increase sensitivity to paclitaxel.

5.2. Personalized Management of Lung Cancer

The tyrosine kinase inhibitor gefitinib (Iressa), which targets the EGFR, is approved for late cases of non-small-cell lung cancer (NSCLC) as a last resort treatment. Most of NSCLC patients do not respond to gefitinib, but about 10% of patients have a rapid and often dramatic clinical response. The molecular mechanisms underlying sensitivity to gefitinib are unknown. It was considered to be a targeted therapy based on the idea that lung cancer might make excess EGFR, and blocking it might slow growth with less toxicity than standard chemotherapy. This growth protein contains a little pocket to capture ATP. Gefitinib apparently targets that pocket, and when the protein is mutated, gefitinib fits inside the pocket much better, blocking ATP and thus inhibiting cancer cell growth. A study from the Massachusetts General Hospital/Dana Farber Cancer Institute (Boston, MA) indicates that response of lung cancer patients to gefitinib is determined by a certain mutation in the *EGFR* gene *(17)*. Eight of nine patients who responded to gefitinib had mutation-containing tumors; seven patients not helped by gefitinib did not. Patients with lung cancer who respond to gefitinib have been reported to have somatic mutations consisting of deletions in exon 19 and in exon 21 of the *EGFR* gene. In addition, a mutation in exon 20 is also associated with acquired resistance to gefitinib in initially gefitinib-sensitive patients.

Laboratory studies of cancer cells show that the mutated receptors are 10 times more sensitive to gefitinib than were normal receptors. The mutations are more common in women, people who had never or not recently smoked, and people who had a subtype called bronchoalveolar cancer. Similar results were obtained in another study where receptor tyrosine kinase genes were sequenced in NSCLC and matched normal tissue *(18)*. EGFR mutations were found in additional lung cancer samples from patients who responded to gefitinib (Eli Lilly & Co's Iressa) therapy and in a lung adenocarcinoma cell line that was hypersensitive to growth inhibition by gefitinib, but not in gefitinib-insensitive tumors or cell lines. These results suggest that EGFR mutations may predict sensitivity to gefitinib. Increased EGFR gene copy number based on FISH analysis is a good predictive marker for response to EGFR inhibitors, stable disease, time to progression, and survival in NSCLC *(19)*. These findings are important as they would enable the development of personalized treatment of cancer. The EGFR Mutation Assay (Genzyme Corporation) detects EGFR mutations in patients with NSCLC that correlate with clinical response to Tarceva® (erlotinib) and IRESSA® (gefitinib). This would enable treatment of responders and even at an earlier stage than the current practice of using it as a last resort. Prospective large-scale clinical studies must identify the most optimal paradigm for selection of patients.

Another drug targeting the EGFR receptor is erlotinib (Tarceva; OSI Pharmaceuticals/Pfizer). A randomized, placebo-controlled, double-blind trial was conducted to determine whether erlotinib prolongs survival in NSCLC after the failure of first-line or second-line chemotherapy *(20)*. Presence or absence of EGFR mutation was not taken into consideration. The results show that erlotinib can prolong survival in patients with NSCLC after first-line or second-line chemotherapy. A clinical trial has compared responsiveness to erlotinib with a placebo for NSLC using tumor biopsy samples from participants in this trial to evaluate EGFR expression immunohistochemically *(21)*. The results indicate that among patients with NSCLC who receive erlotinib, the presence of an EGFR mutation may increase responsiveness to the agent, but it is not indicative of a survival benefit.

5.3. Personalized Management of Prostate Cancer

Currently, measurement of serum PSA is the most useful biomarker for early detection, clinical staging, and monitoring of therapy. However, although on average, men with prostate cancer have higher levels of PSA than healthy men or those with benign prostate diseases, there is a wide variation in levels throughout the population, which leads to false positives and unnecessary biopsies. There is need for a better method of diagnosis and monitoring the course of disease. Improvement in understanding of molecular pathology, earlier diagnosis, and ability to monitor the course of disease in response to treatment would facilitate the development of personalized management of cancer of the prostate.

Significant and widespread differences in gene expression patterns exist between benign and malignant growth of the prostate gland. Gene expression analysis of prostate tissues should help to disclose the molecular mechanisms underlying prostate malignant growth and identify molecular markers for diagnostic, prognostic, and therapeutic use. Alternative proteomic-based approaches including use of LCM enable the identification of protein markers in the actual premalignant and frankly malignant epithelium. Quantitative as well as qualitative proteomic measurement of normal and neoplastic prostate cells is an important approach that will complement genomic DNA and gene expression analyses. The National Cancer Institute Prostate Group has been studying protein profiles of prostate cancer using tissue microdissection and two protein analysis methods: 2D PAGE and SELDI ProteinChip MS technology. A protein fingerprinting technique can be used on serum samples to help accurately distinguish between cancer of the prostate, benign prostate hyperplasia, and healthy tissue.

Id protein family, a group of basic helix-loop-helix transcription factors, has been shown to be involved in carcinogenesis and a prognostic marker in several types of human cancers. A study has examined the expressions of four Id proteins, Id-1, Id-2, Id-3, and Id-4, in clinical prostate cancer specimens as well as nodular hyperplasia specimens by IHC *(22)*. The results indicate that these Id proteins may play a positive role in the development of prostate cancer. Differential Id protein expressions may be a useful marker for poor prognosis, and Id-4 may be a potential prognostic marker for distant metastasis.

Although molecular profiling of cancer at the transcript level has become routine, large-scale analysis of proteomic alterations during cancer progression is been a more daunting task. High-throughput immunoblotting has been used to interrogate tissue extracts derived from prostate cancer *(23)*. This approach has identified numerous proteins that are altered in prostate cancer relative to benign prostate. An integrative analysis of this compendium of proteomic alterations and transcriptomic data revealed only 48–64% concordance between protein and transcript levels. However, differential proteomic alterations between metastatic and clinically localized prostate cancer served as predictors of clinical outcome in prostate cancer.

5.4. Proteomics of Brain Cancer

Protein biomarkers of brain tumors have potential clinical usefulness for predicting efficacy of anticancer agents. In one proteomic study, surgical samples of human gliomas were analyzed with two-dimensional gel electrophoresis (2DE) and MS, and in vitro chemosensitivities to various anticancer agents (e.g., cyclophosphamide, nimustine, cisplatin, cytosine arabinoside, mitomycin C, adriamycin, etoposide, vincristine, and paclitaxel) were measured by flow cytometric detection of apoptosis *(24)*. Proteins that significantly affected the in vitro chemosensitivity to each category of anticancer agents were identified. Many of the proteins that correlated with chemoresistance were categorized into the signal transduction proteins including the G-proteins. This study showed that the proteome analysis using 2DE could provide a list of proteins that may be the potential predictive markers for chemosensitivity in human gliomas. They can also be direct and rational targets for anticancer therapy and be used for sensitization to the conventional chemotherapeutic regimens.

6. CONCLUSIONS

Proteomic technologies have contributed considerably to an understanding of molecular biology of cancer. Discovery of biomarkers will be the basis of tests for early diagnosis of cancer and monitoring of therapy. Proteomics is facilitating the integration of diagnostics and therapeutics and fulfilling many of the requirements for personalized therapy of cancer, which is already gaining clinical recognition. Several new technologies such as nanobiotechnologies are contributing to the development of oncoproteomics and will facilitate the development of personalized management of cancer.

REFERENCES

1. Jain KK. Personalized Medicine: Scientific and Commercial Aspects. Basel, Jain PharmaBiotech Publications, 2007:1–582.
2. Jain KK. Role of pharmacoproteomics in the development of personalized medicine. Pharmacogenomics 2004;5:239–42.
3. Jain KK. Proteomics: Technologies, Markets and Companies. Basel, Jain PharmaBiotech Publications, 2007:1–564.
4. Nettikadan S, Radke K, Johnson J, et al. Detection and quantification of protein biomarkers from fewer than ten cells. Mol Cell Proteomics 2006;5:895–901.
5. Wulfkuhle JD, Liotta LA, Petricoin EF. Proteomic applications for the early detection of cancer. Nat Rev Cancer 2003;3:267–275.
6. Wiesner A. Detection of tumor markers with ProteinChip® technology. Curr Pharm Biotechnol 2004;5:45–67.

7. Miyata T, Jige M, Nakaminami T, Uragami T. Tumor marker-responsive behavior of gels prepared by biomolecular imprinting. Proc Natl Acad Sci USA 2006;103:1190–3.
8. Samoylova TI, Morrison NE, Globa LP, Cox NR. Peptide phage display: opportunities for development of personalized anti-cancer strategies. Anticancer Agents Med Chem 2006;6:9–17.
9. Suber RL, Flanders VL, Campa MJ, Patz EF Jr. Accentuation of differentially expressed proteins using phage technology. Anal Biochem 2004;333:351–7.
10. Da Cruz S, Parone PA, Martinou JC. Building the mitochondrial proteome. Expert Rev Proteomics 2005;2:541–51.
11. Fong S, Debs RJ, Desprez P-Y, et al. Id genes and proteins as promising targets in cancer therapy. Trends Mol Med 2004;10:387–392.
12. Chaurand P, Sanders ME, Jensen RA, Caprioli RM. Proteomics in diagnostic pathology: profiling and imaging proteins directly in tissue sections. Am J Pathol 2004;165:1057–68.
13. Brown RE. Morphoproteomics: exposing protein circuitries in tumors to identify potential therapeutic targets in cancer patients. Expert Rev Proteomics 2005;2: 337–48.
14. Jain KK. Nanobiotechnology: Applications, Markets and Companies. Basel, Jain PharmaBiotech Publications, 2007:1–674.
15. Venne K, Bonneil E, Eng K, Thibault P. Enhanced sensitivity in proteomics analyses using nanoLC–MS and high-field asymmetry waveform ion mobility mass spectrometry. Anal Chem 2005;77:2176–86.
16. Rouzier R, Rajan R, Wagner P, et al. Microtubule-associated protein tau: a marker of paclitaxel sensitivity in breast cancer. Proc Natl Acad Sci USA 2005;102: 8315–20.
17. Lynch TJ, Bell DW, Sordella R, et al. Activating mutations in the epidermal growth factor receptor underlying responsiveness of non-small-cell lung cancer to gefitinib. N Engl J Med 2004;350:2129–39.
18. Paez JG, Janne PA, Lee JC, et al. EGFR mutations in lung cancer: correlation with clinical response to gefitinib therapy. Science 2004;304:1497–500.
19. Hirsch FR, Witta S. Biomarkers for prediction of sensitivity to EGFR inhibitors in non-small cell lung cancer. Curr Opin Oncol 2005;17:118–22.
20. Shepherd FA, Pereira JR, Ciuleanu T, et al. Erlotinib in previously treated non-small-cell lung cancer. N Engl J Med 2005;353:123–32.
21. Tsao MS, Sakurada A, Cutz JC, et al. Erlotinib in lung cancer ? molecular and clinical predictors of outcome. N Engl J Med 2005; 353:133–44.
22. Yuen HF, Chua CW, Chan YP, et al. Id proteins expression in prostate cancer: high-level expression of Id-4 in primary prostate cancer is associated with development of metastases. Mod Pathol 2006;19:931–41.
23. Varambally S, Yu J, Laxman B, et al. Integrative genomic and proteomic analysis of prostate cancer reveals signatures of metastatic progression. Cancer Cell 2005;8:393–406.
24. Iwadate Y, Sakaida T, Saegusa T, et al. Proteome-based identification of molecular markers predicting chemosensitivity to each category of anticancer agents in human gliomas. Int J Oncol 2005;26:993–8.

5 Application of Serum and Tissue Proteomics to Understand and Detect Solid Tumors

Christina M. Annunziata, Dana M. Roque, Nilofer Azad, and Elise C. Kohn

CONTENTS

1. INTRODUCTION
2. MOLECULAR PROTEOMIC TECHNIQUES FOR CLINICAL TRIALS
3. INCORPORATION OF MOLECULAR PROTEOMICS INTO CLINICAL TRIALS
4. CONCLUSIONS

SUMMARY

Proteomics embodies a diverse set of platforms devoted to the study of protein structure and function. Mass spectrometry and protein microarrays enable detailed analysis of a diverse set of proteins and their isoforms at the molecular level. These new, high-throughput, and comprehensive proteomic studies can be useful in advancing our understanding of disease processes and have wide potential for integration within clinical trials. Clinical trials incorporating proteomic-based endpoints are geared toward biomarker development, pathway discovery, and target validation. Screening the proteome has the potential to discover a protein biomarker or

From: *Cancer Drug Discovery and Development
Cancer Proteomics: From Bench to Bedside*
Edited by: S. S. Daoud © Humana Press Inc., Totowa, NJ

a signature of markers to predict disease presence, and we have initiated a multi-center clinical trial with the goal of defining a proteomic signature that can predict relapse of ovarian cancer. Serum and tissue proteomics reflect information from the tumor as well as from the tumor microenvironment, as demonstrated in our two recent clinical trials employing these analyses to evaluate effects of targeted agents. Proteomic techniques strengthen the link between bench and bedside by testing existing hypotheses about a disease state or treatment and validating them in the clinical setting. Defining signaling pathways and protein regulatory events in the clinical venue may thus lead to generation of new hypotheses and subsequently to novel treatment strategies.

Key Words: Proteomics; clinical trial; solid tumor; mass spectrometry; tissue microarray

1. INTRODUCTION

The field of proteomics embodies a diverse set of platforms devoted to the study of protein structure and function. These techniques possess the capacity to capture the complexity of the human proteome, which has been approximated to contain at least 10,000,000 unique peptides as a result of alternative splicing, genomic amplification, and intra-/extra-cellular cleavage events *(1)* While the study of single-nucleotide polymorphisms *(2)* and mRNA patterns via cDNA microarrays *(3)* have been shown to be useful in the prediction of disease outcomes, a mere understanding of the genome and transcriptome do not fully unravel the mechanisms underlying pathological processes. It has been approximated that the correlation between mRNA and protein expression is often no greater than 40% *(3)*. Furthermore, information regarding quaternary structure, secretory patterns, and post-translational modifications such as phosphorylation cannot be derived from knowledge of amino acid sequences alone *(4)*.

Proteomic studies are thus crucial to the understanding of disease processes and have enormous potential for integration within clinical trials. Modern proteomic techniques have been employed to characterize many diverse pathologic states including cardiac hypertrophy *(5)*, transplantation rejection *(6)*, diabetic nephropathy/retinopathy *(7,8)*, growth hormone disorders *(9)*, as well as tumor biology of the prostate, gastrointestinal system, breast, head/neck, and ovary *(10–18)*.

Proteomic-based clinical trials are involved in biomarker development, pathway discovery, and target validation. The massive influx of proteomics data into clinically oriented literature in recent years has generated excitement within the scientific community and has rejuvenated interest in translational science. Clinical trials incorporating proteomic-based endpoints are useful for biomarker development, pathway discovery, and target validation. The need for biomarker discovery using proteomic techniques cannot be understated. Biomarkers have the potential to improve detection of disease, classify histologic subtypes, predict progression, and forecast responsiveness to a particular treatment *(19,20)*. There are currently only eight Food and Drug Administration-approved serum tumor markers for the detection and monitoring of malignancies *(21)*. Unfortunately, these markers all perform with sensitivities and specificities suboptimal for detecting new cancers in the general population. For example, serum CA-125 has been approved for monitoring treatment or recurrence in a woman with previously diagnosed ovarian cancer, but an elevated serum CA-125 (>35 U/ml) is detectable in only 50–60% of patients with newly diagnosed stage I ovarian cancer and may be influenced by benign ovarian and other processes *(22)*. Similar problems exist with carcinoembryonic antigen for colorectal cancer, prostate-specific antigen for prostate cancer, CA 15-3 and CA-27.9 for breast cancer, CA 19.9 for pancreatic cancer, alpha fetoprotein for liver cancer, and nuclear matrix protein 22 for bladder cancer *(23)*. The majority of human diseases still have no identifiable biomarkers. Recent efforts have concentrated on applying proteomic techniques to identify novel serum biomarkers *(24–26)*, correlate kinase pathway activation with prognosis *(27)*, and to distinguish tumor from normal tissue *(28)*.

Proteomic-based studies to identify aberrancies of molecular pathways in disease states are also critical to the development of novel pharmaceuticals. Presently, the mechanistic basis of therapeutic agents is surprisingly homogenous, with guanine nucleotide protein-coupled receptors constituting nearly 45% of all protein targets. Even if only 1–2% of all proteins were found to be useful as loci for intervention, an additional 5000–10,000 candidates remain unexploited *(29)*. The identification of signaling nodes representative of interaction points between molecular pathways is vital to the development of a more intricate comprehension of signal transduction. It is crucial to target these sites in signaling communication wherein the clinically optimal interruption can occur. The importance of knowing the complexities of these communications is highlighted by the fact that targeted agents are not uniformly focused in their actual actions in vivo. Sorafenib, for example, was designed to target Raf kinase, but shows clinically relevant activity against vascular endothelial growth factor receptor-2

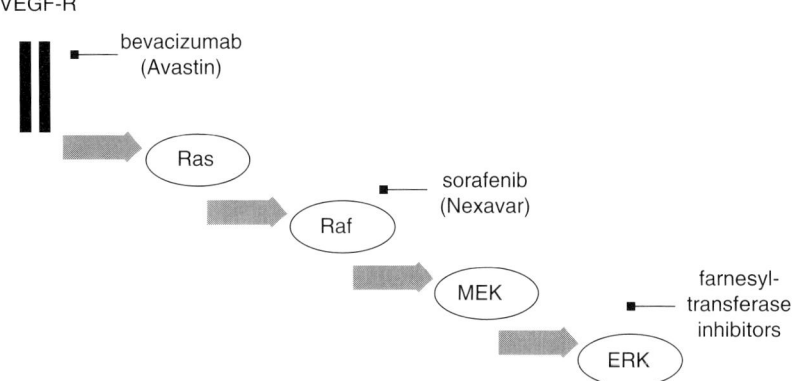

Fig. 1. Combinatorial therapies targeting multiple points within the same pathway have the potential to induce therapeutic synergism and reduce single-agent toxicities. Illustrated here is the proposed serial inhibition by bevacizumab and sorafenib.

(VEGFR-2) as well. Elucidation of compound pathways has fueled an interest in the design of combinatorial therapies that would target different points within the same pathway in order to induce therapeutic synergism and reduce single-agent toxicity (Fig. 1).

Finally, proteomic applications are essential for validation of a drug's effect on its designated target. Such validation is achieved when a compound selective for a disease target shows efficacy in the affected population as a result of specific changes in the intended molecular endpoints *(30)*. Currently, a number of monoclonal antibodies and small molecule therapies have entered this stage of development at the level of single-agent or combinatorial therapy *(31)*, exemplified by bevacizumab (anti-VEGF) *(32)*, cetuximab [anti-epidermal growth factor receptor, (EGFR)] *(33,34)*, trastuzumab (anti-her2) *(35,36)*, gefitinib (EGFR tyrosine kinase inhibitor) *(37–40)*, and imatinib (abl/c-kit receptor tyrosine kinase inhibitor) *(41–43)*.

2. MOLECULAR PROTEOMIC TECHNIQUES FOR CLINICAL TRIALS

2.1. Overview

The understanding of protein structure and function has evolved substantially over the past 30 years, catalyzed in part by advances in molecular biology at the genomic level *(44)*. Recently, technical innovations have focused largely on two main areas: first, increasing the sensitivity of protein detection *(1)* to thereby reduce sample volumes required for meaningful protein analysis, and second, accelerating instrumentation for high-throughput processing. Techniques available for the study of protein chemistry in the clinical setting have thus evolved to fall under two broad headings: targeted analyses and high-throughput analyses.

2.2. Targeted Analyses

2.2.1. IMMUNOHISTOCHEMISTRY

Immunohistochemistry (IHC) predated the era of proteomics but emphasizes the importance of protein analysis in the clinical setting. IHC is a technique for the visualization of specific antigens in situ, first reported by Coons and colleagues in 1942. Sample integrity is preserved through sample flash freezing or fixation, the latter done most commonly with neutral-buffered formalin followed by paraffin embedding. Antigen retrieval is accomplished through heating or enzymatic treatment prior to interrogation with a chromagenically or fluorescently labeled antibody. The greatest strength of IHC lies in its preserved tissue organization, allowing cellular and subcellular localization of proteins. Unfortunately, it is relatively low-throughput, requires individual optimization for each antibody of interest, and has day-to-day variation that may alter intensity of stain that is assessed by many investigators. Additionally, formalin fixation results in cross-linkage of antigens that may mask some protein epitopes *(45)*. IHC has broad clinical applications including diagnostics and assessment of prognostic markers. For example, the presence of hormone receptors in breast cancer suggests responsiveness to anti-hormone treatment, or the level of mutated c-kit in gastrointestinal stromal tumor allows treatment with imatinib.

2.2.2. GEL ELECTROPHORESIS AND IMMUNOBLOTTING

Also a prelude to the proteomic era, one-dimensional denaturing gel electrophoresis has been a standard technique for the semi-quantitative study of protein expression since its introduction more than 30 years ago by Laemmli *(46)*. Proteins are resolved on the basis of charge in isoelectric focusing and molecular weight in standard approaches. Proteins transferred to a nitrocellulose or polyvinylidene fluoride membrane can be identified by immunoblot analysis with specific antibody probes. Two-dimensional gel electrophoresis (2DE) uses separation in both directions; differential in-gel electrophoresis represents a further refinement in which samples are tagged with fluorescent labels prior to electrophoretic processing. Differential expression from the same gel allows individual proteins of interest to be removed from the gel and subjected to identification by mass spectrometry (MS) and protein sequencing *(47)*.

Protein electrophoresis as described cannot be applied to very small clinical samples, cannot resolve proteins which are present in minute concentrations, and cannot adequately identify very low molecular weight proteins, under the range of 5–6 KDa *(48)*. 2DE has been applied to archived samples from past clinical trials involving the study of neurologic disorders *(49)*, atherosclerosis *(50)*, and ovarian cancer *(51)*. While a typical 2DE analysis is capable of separating 3000 spots per experiment *(12)*, the technique remains labor-intensive and not applicable to small clinical samples or high-throughput analysis.

2.3. High-Throughput Analyses

2.3.1. PROTEIN ARRAYS

Two broad protein microarray platforms currently exist: reverse phase and forward phase. The reverse-phase array (Fig. 2A) was described initially by Liotta and colleagues *(52)*. Cells taken from culture or extracted from tissues of interest, often through laser capture microdissection *(53)*, are lysed and arrayed on a nitrocellulose-coated slide. Printed arrays are probed with immunoblot-validated antibodies, subjected to a signal amplification process, and detected using a colorimetric or fluorescent dye. Signal intensities representative of protein concentration are quantified in pixels and used for comparison across samples. In the forward-phase array (Fig. 2B), antibodies are pre-bound to nitrocellulose as bait for cellular lysates, which are then detected through the application of a labeled antibody *(54,55)*. Alternatively, the forward-phase array may be performed with lysates labeled with biotin and then detected with avidin-coupled fluorescent reagent, or the lysate may be directly labeled with the fluorescent (Cy3 or Cy5) dye *(56)*.

Advantages of the protein microarray are twofold. First, protein microarrays attain enhanced sensitivity for protein detection relative to IHC or immunoblot, thus reducing sample volume required from patient biopsies. Secondly, they generate data in a high-throughput manner as commercially available arrayers now have the capacity to create protein chips containing more than 13,000 spots per standard microscope slide. Development of this technology continues; results

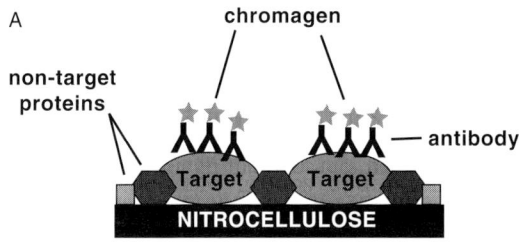

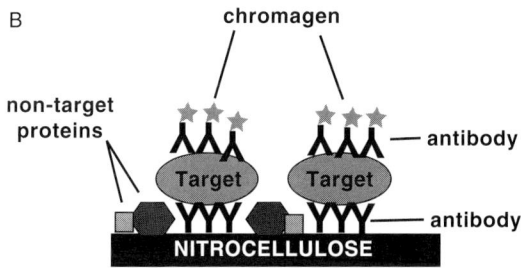

Fig. 2. Illustration of reverse-phase (A) and forward-phase (B) protein microarray formats.

may be complicated by intra- and inter-run variability and there remains a need to improve and standardize the algorithms available for the analysis of the resultant data. We have optimized quality control of tissue handling, removing one source of variability. Ongoing quality control studies at our institution and others are rapidly improving the sensitivity, specificity, reliability, and validity of this technique with the use of clinical samples.

Both forward- and reverse-phase tissue lysate arrays have been applied retrospectively to patient samples. Protein microarrays have been used to identify up- or down-regulation of proteins in hepatocellular carcinoma versus normal liver *(57)*, characterize the influence of microenvironment on the differential characteristics of primary colorectal tumors versus their hepatic metastases *(28)*, describe changes in Ak mouse thymoma (AKT) signaling with progression of prostate cancer *(52)*, and examine phosphorylation pathways in metastatic ovarian carcinoma *(58)*. Studies at our institution have now progressed to incorporate reverse-phase arrays in prospective trials in an effort to advance the era of personalized molecular medicine.

2.3.2. MASS SPECTROMETRY

MS was developed in the early 1900s as a tool for the identification of proteins. This technology had remained relatively dormant in the clinical literature until the recent availability of high-resolution mass spectrometers and the sequencing of the human genome. In matrix-assisted laser desorption/ionization time of flight (MALDI-TOF) MS, biological analytes, purified of inorganic salts and other contaminants by chromatographic or electrophoretic means, are prepared in a light-absorbent matrix and then deposited on a solid surface. The laser-applied energy produces a phase transition from solid to gas wherein proteins are ionized into peptide fragments. The peptides are then subjected to an electric field under vacuum, which separates components on the basis of mass. The resultant output is displayed as a spectrum of peaks plotted by intensity versus mass-to-charge (m/z) ratio.

Surface-enhanced laser desorption/ionization (SELDI)-TOF is a refinement of MALDI introduced in 1993 by Hutchens and Yip in which solid-phase purification is accomplished on a chip selective for proteins based on chemical affinities *(59)*. Samples may be analyzed on multiple chip surfaces and the data compared or compiled, adding complexity but potentially more information. Relative to MALDI, this platform is less expensive and less labor-intensive, and allows the analysis of whole proteins rather than peptide fragments, though SELDI-TOF spectra generally demonstrate lower resolution (Fig. 3).

MS techniques have potential for use in clinical trials by offering a high-throughput algorithm with detection limits in the femtomolar range. MS platforms are ideal for the global analysis of protein content in biological specimens, and there is growing interest in using MS to explore the low molecular weight range of the proteome (<20 kDa) as a source of novel biomarkers *(24)*. Like all technologies, MS analyses can suffer from the biases of day-to-day instrument variation obviated by frequent calibration, normalization of peak intensities, and the use of internal standards *(48)*. The study of low abundance proteins may be confounded by the presence of dominant species such as albumin, transferrin, and immunoglobulins which represent as much as 80% of serum proteins. On the other hand, such proteins may be exploited as carriers to amplify low abundance proteins to a level detectable by MS *(60)*.

Many groups have focused on the development of serum-based screening tests for occult cancers. In addition to the initial and sometimes controversial report of its use in ovarian cancer *(24)*, MS

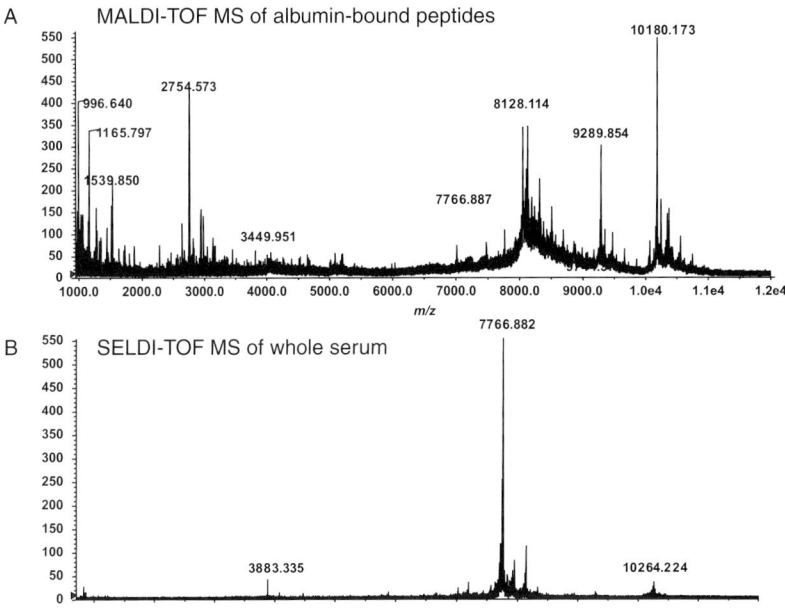

Fig. 3. Sensitivity of mass spectroscopic output is greatly influenced by sample preparation and ionization platform. Spectra obtained from samples enriched for albumin-bound peptides as analyzed by matrix-assisted laser desorption/ionization time of flight (MALDI-TOF) and non-enriched samples analyzed by surfaced-enhanced laser desorption/ionization time-of-flight (SELDI-TOF) are shown in panels A and B, respectively. Normalized signal intensity and mass to charge ratio (m/z) are represented on y-axis and x-axis, respectively.

techniques have yielded panels of proteins that discriminate non-invasive prostate cancer from benign prostatic hypertrophy *(61)* and detect breast and ovarian cancers with reported specificities and sensitivities of 90–100% *(62–64)*. MS data from solid tissue samples have been developed toward discrimination of benign and malignant breast disease *(12)*, as well as cervical cancer from normal cervix tissue *(65)*. MS analyses have been employed to examine differences in protein expression between germ-line BRCA-1 mutation-positive women who do and do not develop cancer *(66)* and to identify candidate biomarkers for pancreatic cancer such as annexin A2 *(62)*. MS has also been performed on fluids other than serum, such as ascites *(67)* and urine *(63)*. Treatment-induced changes in urinary SELDI profiles have been observed in a phase I study of intravesical suramin for transitional cell bladder cancer *(68)*.

Among the challenges of molecular clinical proteomics are the complexity of a dynamic proteome. Though the methods have generated excitement in the area of clinical proteomics, progress has been met with justified skepticism for a number of reasons. First, the techniques and instrumentation for clinical proteomics are continuously evolving, making rigorous quality control and quality assurance difficult to achieve *(69,70)*. The scientific community is beginning to address this responsibility by establishing formal guidelines for the identification and study of tumor markers as has been done for the use and reporting of gene expression arrays *(71)*. Second, the large datasets generated from assaying the proteome require sophisticated analytical algorithms for processing *(69)*, leading to questions of prospectively determining sample size based upon statistical power. Our initial work in ovarian cancer was powered with a dataset of 50 cases which gave 96% power at the alpha = 0.05 level to reject an 80% sensitivity or specificity in favor of a true value of 95%, using an exact test for single proportions, with cut-off points for rejection based on the cumulative binomial distribution *(24)*. This small number and limited statistical power was selected for proof of concept. More stringent power is needed in order to develop confidence for clinical application. Statistical overfitting is a concern that must be considered in powering these types of studies. The large data streams obtained from each case leaves a risk of statistical inaccuracy by overfitting if >1000 potential markers are analyzed simultaneously *(70)*. Many current MS biomarker studies are designed simply to characterize samples, with the hopes of uncovering significant and applicable trends, rather than planned and powered for a defined hypothesis. Third, the human proteome is dynamic and exquisitely sensitive to inflammatory processes, age, trauma, diurnal variations, fasting versus non-fasting states, quality of sample acquisition and processing, and other unidentified variables *(72)*. Clinical specimens consist of thousands of proteins, some of which may have several isoforms *(73)* and may be altered by collection methodology. These variables are further complicated by collection variables, making quality control and standard operating procedure definitions critical to any sample collection. The variability in reproducing endpoints is likely due to this complexity of biological materials.

Serum and tissue proteomics reflect information from the tumor as well as from the tumor microenvironment. Tumor specimens are heterogeneous tissues containing red blood cells, inflammatory cells, parenchymal cells of non-tumor origin clonal variants, and areas of necrosis, and like serum must be handled in careful and consistent ways to minimize collection bias and sample degradation. Thus, it is clear that the integration and cross-validation of the various proteomic techniques will be required to strengthen proteomics-based research. The scientific community may then move to replace skepticism surrounding proteomic conclusions with an interest in adequate quality control and novel clinical trial design.

3. INCORPORATION OF MOLECULAR PROTEOMICS INTO CLINICAL TRIALS

3.1. Clinical Trial Endpoints

Traditional endpoints of a drug or intervention in clinical trials are toxicity (Phase I), efficacy (Phase II), and survival (Phase III). The introduction of proteomic investigations into the trial(s) may allow greater depth of analysis into the cause and effect relationship between the drug and observed outcomes. Rather than simply answer the question of drug effect, proteomics may allow the exploration of multiple parameters surrounding the effect or lack thereof, for example:

- Did the drug affect its intended target or other off-targets?
- Did it act in the tumor itself or the surrounding stroma?
- Are downstream pathways of the intended target intact or deranged in the tumor or stroma?
- Is the magnitude of targeted effect adequate at a level below maximum-tolerated dose?

Proteomic investigations extend translational inquiry to a broader scale, by exploring a large fraction of a patient's expressed proteins. Both target validation and biomarker discovery are potential uses of these high-throughput methods. Proteomic investigations are therefore applicable to treatment trials and screening trials alike.

3.2. Clinical Trial Design

3.2.1. SCREENING TRIALS: BIOMARKER DISCOVERY

Screening the proteome has the potential to discover a protein biomarker or a signature of markers to predict disease presence. For a biomarker or signature to be of value, it should have sufficient sensitivity and specificity, come from an easily sampled source such as blood, and trigger an effective intervention. The sensitivity and specificity of a defined biomarker is determined by two sets of patients.

Those without active disease—either in remission, at risk of developing the disease, or unaffected—provide a negative control for specificity. The second group, those affected by the disease, determines the sensitivity of the defined biomarker. From both groups, the predictive values of the biomarker can be calculated. Power of the study is critical and depends upon the stringency of the anticipated sensitivity and specificity as well as the complexity of the biomarker. A single biomarker of disease could theoretically exist and would reveal itself at the time of disease occurrence or relapse. More likely, however, a disease will cause a series of alterations in serum proteins. A signature of cancer might be evident upon comparing the composition of the serum proteome at the time of illness with that in the remission state or in unaffected patients. Proteomic technology has been applied to serum samples from patients with prostate cancer *(25)*, colon cancer *(74)*, and ovarian cancer *(75)* to detect proteomic signatures that distinguish patients with cancer from those without cancer. A biomarker containing many features coming from a large data stream, such as those identified using MS technologies described above, requires a higher power because of the risk of overfitting.

A clinical trial design for proteomic-based screening incorporates quality-controlled sample acquisition and storage. Sampling times must be logical and useful against the objective of the study such that they identify changes in target events during the period of observation. In a screening trial, no detectable tumor may exist; serial sampling of patients' serum is required in such cases. Serial physical examinations and/or imaging are the classical tools for diagnosing the presence of disease. These methods are inefficient screening tools due to low sensitivity and high cost. Ideally, a serum proteomic approach to diagnosis might be easier and more cost effective if it is sensitive enough to detect disease at an early time point and specific enough to differentiate benign from malignant conditions.

Consistent and reliable disease identification should also lead to a clinical intervention that, in the long run, should make a difference in intervention and outcome. For example, screening for cervix cancer identifies premalignant disease that when treated improves survival by preventing progression to advanced incurable disease.

We have initiated a multi-center clinical trial with the goal of defining a proteomic signature that can predict relapse of ovarian cancer. This objective was designed in part to develop a signature of minimal residual disease. It is hypothesized that this minimal residual disease signature might also be applicable to early diagnosis and will be applied to serum samples of unaffected or high-risk patients in an exploratory fashion. Our trial is enrolling women in first clinical remission from treatment for newly diagnosed stage III or IV ovarian cancer. They are eligible to enroll within 3 months of completing adjuvant chemotherapy. Women will be followed clinically with physical examination and blood work every 3 months, and CT scan every 6 months while on study. Serum samples will be collected every 3 months for later proteomic analysis. The primary endpoint of the clinical trial is to identify a proteomic signature in the serum that can predict relapse of ovarian cancer before the recurrent tumor becomes evident on physical exam or imaging study. The study is powered at 300 patients using the same statistical argument described above for the training set of 50 recurrences and 50 patients with no evidence of disease. The remainder of the cohort will yield at least another 50 affected and 50 remitted patients for independent validation of the signature sets developed by those using this repository resource to develop proteomic signatures.

3.2.2. TREATMENT TRIALS: TARGET VALIDATION

In a clinical trial aimed at cancer treatment, the ideal tissue to analyze for a drug's targeted effect is the tumor itself and its surrounding stroma. Skin or buccal mucosa have been considered for use as surrogate tissues when examining a specific effect of an investigational agent but have not proven to reflect accurately the results in the tumor itself *(76)*. Tumor biopsies, therefore, should be performed within a clinical trial when feasible and safe. Serum samples can extend analysis to the secreted proteome and can provide further validation of pharmacologic outcome as well as new target discovery.

We recently completed two clinical trials applying tissue and serum proteomics to evaluate effects of targeted agents. The first clinical trial assessed the targeted activity of the EGFR inhibitor gefitinib in 24 patients with ovarian cancer. Percutaneous 16- to 18-g core needle tumor biopsies were obtained prior to beginning therapy and after 4 weeks of gefitinib treatment. Tumor and stromal cells were separated using laser capture microdissection and analyzed separately by tissue lysate arrays. Arrayed samples were analyzed for the levels of EGFR and its activated (phosphorylated) forms, as well as downstream signaling molecules AKT and ERK, and their activated forms. Results confirmed gefitinib's effectiveness at inhibiting its intended target and downstream signaling. No clinical benefit was seen, suggesting that target inhibition was present but not essential for tumor growth, the level of inhibition was insufficient to completely prevent receptor signaling, and/or the receptor function was unrelated to the malignancy at this advanced and recurrent state of disease. Sample statistics from the gefitinib study show a relationship for signaling parameter and clinical toxicity by grade and by trend of toxicity. Increasing EGFR, AKT, p-ERK, and p-EGFR moieties in tumor after treatment were statistically significantly associated with increasing overall toxicity ($p \leq 0.05$), gastrointestinal toxicity ($p < 0.05$), and skin toxicity ($p = 0.029$) *(77)*. We also studied the signal transduction inhibitor, imatinib, in women with relapsed ovarian cancer. Using a similar biopsy and array paradigm, we found association of biochemical modulation of c-kit and EGFR expression with phospho-c-kit and EGFR correlating with nausea and vomiting, and post-treatment levels of EGFR and PDGFR linked to fatigue *(78)*. A surprising number of patients, approximately 80%, had evidence of c-kit and p-c-kit expression in their tumor and/or stroma, compared with the 12% predicted by IHC *(79,80)*. This illustrates the strength of the tissue lysate array. These clinical trials demonstrated the ability to incorporate invasive sample collection and detailed biochemical analysis to confirm target modulation and to correlate the biochemical proteomic events with clinical and toxicity events.

3.2.3. COMBINATION THERAPY, INDIVIDUAL EFFECTS

Therapy targeted at sequential points in a pathway critical for cell survival has the potential to effectively eliminate the cancer cell. Knowledge of signaling pathways in vivo gained from proteomic investigations can inspire logical combinations of targeted agents. When agents are combined, individual effects may be assessed if the clinical trial is designed to randomize start with single-agent therapy for the first cycle followed by combination therapy beginning with the subsequent cycle. We are currently treating patients on a clinical trial evaluating the combination of bevacizumab and sorafenib, two agents directed at VEGFR-2 signaling. The complexity of the dual signal interruption is difficult to analyze using traditional endpoints. We are again using prospectively planned serial biopsies to allow a broad biochemical assessment of target modulation by the agents. In addition to the focused endpoints, the requirement for small amounts of tissue for the lysate array leaves additional remaining biopsy material that may be applied to further array biochemical endpoints or cut for more traditional immunohistochemical studies to complement the array results.

4. CONCLUSIONS

Proteomics increases the capacity of clinical inquiry to answer specific biochemical questions regarding disease etiology and effects of clinical intervention. Advances in serum proteomics are allowing a detailed analysis of the vast secreted and circulating proteome. Data mining from this resource has the potential to reveal many disease-specific peptide combination signatures. Tissue proteomics permits in vivo confirmation of an intervention's specific intended effects and exploration of related downstream events. It can be used for discovery and proof of concept studies. The high throughput nature of both types of proteomics technologies is ideal for the rapid analytic platforms. Furthermore, MS technology is currently used in many clinical arenas and is ready to be applied in the clinical trial setting. These techniques strengthen the link between bench and bedside by testing existing hypotheses about a disease state or treatment and validating them in the clinical setting. Defining signaling pathways and protein regulatory events in the clinical venue may thus lead to generation of new hypotheses and subsequently to novel treatment strategies.

ACKNOWLEDGMENTS

This research was supported in part by the Intramural Research Program of the NIH, National Cancer Institute, Center for Cancer Research.

REFERENCES

1. Hortin GL, Shen RF, Martin BM, Remaley AT. Diverse range of small peptides associated with high-density lipoprotein. Biochem Biophys Res Commun 2006;340(3):909–15.
2. Engle LJ, Simpson CL, Landers JE. Using high-throughput SNP technologies to study cancer. Oncogene 2006;25(11):1594–601.
3. van de Vijver MJ, He YD, van't Veer LJ, et al. A gene-expression signature as a predictor of survival in breast cancer. N Engl J Med 2002;347(25):1999–2009.
4. Zhang ZY. Protein tyrosine phosphatases: structure and function, substrate specificity, and inhibitor development. Annu Rev Pharmacol Toxicol 2002;42:209–34.
5. Faber MJ, Agnetti G, Bezstarosti K, et al. Recent developments in proteomics: implications for the study of cardiac hypertrophy and failure. Cell Biochem Biophys 2006;44(1):11–29.
6. Traum AZ, Schachter AD. Transplantation proteomics. Pediatr Transplant 2005;9(6):700–11.
7. Ahn BY, Song ES, Cho YJ, Kwon OW, Kim JK, Lee NG. Identification of an anti-aldolase autoantibody as a diagnostic marker for diabetic retinopathy by immunoproteomic analysis. Proteomics 2006;6(4):1200–9.

8. Merchant ML, Klein JB. Proteomics and diabetic nephropathy. Curr Diab Rep 2005;5(6):464–9.
9. Chung LP, Clifford D, Buckley M, Baxter RC. Novel biomarkers of human growth hormone action from serum proteomic profiling using protein chip mass spectrometry. J Clin Endocrinol Metab 2006;91(2):671–7.
10. Becker S, Cazares LH, Watson P, et al. Surfaced-enhanced laser desorption/ionization time-of-flight (SELDI-TOF) differentiation of serum protein profiles of BRCA-1 and sporadic breast cancer. Ann Surg Oncol 2004;11(10): 907–14.
11. Dey P. Urinary markers of bladder carcinoma. Clin Chim Acta 2004;340(1–2): 57–65.
12. Dwek MV, Alaiya AA. Proteome analysis enables separate clustering of normal breast, benign breast and breast cancer tissues. Br J Cancer 2003;89(2):305–7.
13. Engel C, Forberg J, Holinski-Feder E, et al. Novel strategy for optimal sequential application of clinical criteria, immunohistochemistry and microsatellite analysis in the diagnosis of hereditary nonpolyposis colorectal cancer. Int J Cancer 2006;118(1):115–22.
14. Hirota S, Isozaki K. Pathology of gastrointestinal stromal tumors. Pathol Int 2006;56(1):1–9.
15. Pawlik TM, Kuerer HM. The evolving role of proteomics in the early detection of breast cancer. Int J Fertil Womens Med 2005;50(5 Pt 1):212–6.
16. Semmes OJ, Malik G, Ward M. Application of mass spectrometry to the discovery of biomarkers for detection of prostate cancer. J Cell Biochem 2006;98(3): 496–503.
17. Beer HL, Jenkins RE, Gutowska-Owsiak D, Pazmany L, Birchall MA, Kitteringham NR. 2-Dimensional gel electrophoresis and MALDI-MS of cystic cervical metastasis from head and neck squamous cell carcinoma. Clin Otolaryngol 2006;31(3):246.
18. Tchabo NE, Liel MS, Kohn EC. Applying proteomics in clinical trials: assessing the potential and practical limitations in ovarian cancer. Am J Pharmacogenomics 2005;5(3):141–8.
19. Alaiya A, Al-Mohanna M, Linder S. Clinical cancer proteomics: promises and pitfalls. J Proteome Res 2005;4(4):1213–22.
20. Liotta LA, Kohn EC, Petricoin EF. Clinical proteomics: personalized molecular medicine. JAMA 2001;286(18):2211–4.
21. Rai AJ, Chan DW. Cancer proteomics: serum diagnostics for tumor marker discovery. Ann N Y Acad Sci 2004;1022:286–94.
22. Rosen DG, Wang L, Atkinson JN, Yu Y, Lu KH, Diamandis EP, Hellstrom I, Mock SC, Liu SC, Bast RC Jr. Potential markers that complement expression of CA125 in epithelial ovarian cancer. Gynecol Oncol 2005; 99:267–77.
23. Hoefner DM. Serum tumor markers. Part I: clinical utility. MLO Med Lab Obs 2005;37(12):20, 22–4.
24. Petricoin EF, Ardekani AM, Hitt BA, et al. Use of proteomic patterns in serum to identify ovarian cancer. Lancet 2002;359(9306):572–7.
25. Petricoin EF, III, Ornstein DK, Paweletz CP, et al. Serum proteomic patterns for detection of prostate cancer. J Natl Cancer Inst 2002;94(20):1576–8.
26. Zhang Z, Bast RC, Jr., Yu Y, et al. Three biomarkers identified from serum proteomic analysis for the detection of early stage ovarian cancer. Cancer Res 2004;64(16):5882–90.
27. Zhang D, Tai LK, Wong LL, Chiu LL, Sethi SK, Koay ES. Proteomic study reveals that proteins involved in metabolic and detoxification pathways are

highly expressed in HER-2/neu-positive breast cancer. Mol Cell Proteomics 2005;4(11):1686–96.
28. Petricoin EF, Bichsel VE, Calvert VS, et al. Mapping molecular networks using proteomics: a vision for patient-tailored combination therapy. J Clin Oncol 2005;23(15):3614–21.
29. Drews J. Drug discovery: a historical perspective. Science 2000;287(5460): 1960–4.
30. Kopec KK, Bozyczko-Coyne D, Williams M. Target identification and validation in drug discovery: the role of proteomics. Biochem Pharmacol 2005;69(8):1133–9.
31. Posadas EM, Simpkins F, Liotta LA, MacDonald C, Kohn EC. Proteomic analysis for the early detection and rational treatment of cancer–realistic hope? Ann Oncol 2005;16(1):16–22.
32. Jubb AM, Hurwitz HI, Bai W, et al. Impact of vascular endothelial growth factor-A expression, thrombospondin-2 expression, and microvessel density on the treatment effect of bevacizumab in metastatic colorectal cancer. J Clin Oncol 2006;24(2):217–27.
33. Robert F, Ezekiel MP, Spencer SA, et al. Phase I study of anti-epidermal growth factor receptor antibody cetuximab in combination with radiation therapy in patients with advanced head and neck cancer. J Clin Oncol 2001;19(13):3234–43.
34. Vincenzi B, Santini D, Russo A, et al. Angiogenesis modifications related with cetuximab plus irinotecan as anticancer treatment in advanced colorectal cancer patients. Ann Oncol 2006;17(5):835–41.
35. Esteva FJ, Valero V, Booser D, et al. Phase II study of weekly docetaxel and trastuzumab for patients with HER-2-overexpressing metastatic breast cancer. J Clin Oncol 2002;20(7):1800–8.
36. Gennari R, Menard S, Fagnoni F, et al. Pilot study of the mechanism of action of preoperative trastuzumab in patients with primary operable breast tumors overexpressing HER2. Clin Cancer Res 2004;10(17):5650–5.
37. Cohen EEW, Kane MA, List MA, et al. Phase II trial of gefitinib 250 mg daily in patients with recurrent and/or metastatic squamous cell carcinoma of the head and neck. Clin Cancer Res 2005;11(23):8418–24.
38. Janmaat ML, Gallegos-Ruiz MI, Rodriguez JA, et al. Predictive factors for outcome in a phase II study of gefitinib in second-line treatment of advanced esophageal cancer patients. J Clin Oncol 2006;24(10):1612–9.
39. Lassman AB, Rossi MR, Razier JR, et al. Molecular study of malignant gliomas treated with epidermal growth factor receptor inhibitors: tissue analysis from North American Brain Tumor Consortium Trials 01–03 and 00–01. Clin Cancer Res 2005;11(21):7841–50.
40. Ogino S, Meyerhardt JA, Cantor M, et al. Molecular alterations in tumors and response to combination chemotherapy with gefitinib for advanced colorectal cancer. Clin Cancer Res 2005;11(18):6650–6.
41. Coleman RL, Broaddus RR, Bodurka DC, et al. Phase II trial of imatinib mesylate in patients with recurrent platinum- and taxane-resistant epithelial ovarian and primary peritoneal cancers. Gynecol Oncol 2006;101(1):126–31.
42. Johnson FM, Krug LM, Tran HT, et al. Phase I studies of imatinib mesylate combined with cisplatin and irinotecan in patients with small cell lung carcinoma. Cancer 2006;106(2):366–74.
43. Vuky J, Isacson C, Fotoohi M, et al. Phase II trial of imatinib (Gleevec (R)) in patients with metastatic renal cell carcinoma. Invest New Drugs 2006;24(1):85–8.

44. Patterson SD, Aebersold RH. Proteomics: the first decade and beyond. Nat Genet 2003;33:311–23.
45. Hua C, Langlet C, Buferne M, Schmittverhulst AM. Selective destruction by formaldehyde fixation of an H-2kb serological determinant involving lysine-89 without loss of T-cell reactivity. Immunogenetics 1985;21(3):227–34.
46. Laemmli UK. Cleavage of structural proteins during assembly of head of bacteriophage T4. Nature 1970;227(5259):680–5.
47. Simpkins F, Czechowicz JA, Liotta L, Kohn EC. SELDI-TOF mass spectrometry for cancer biomarker discovery and serum proteomic diagnostics. Pharmacogenomics 2005;6(6):647–53.
48. Winters M, Lowenthal M, Feldman A, Liotta L. The future of cancer diagnostics: proteomics, immunopreoteomics, and beyond. In: Detrick B HR, Folds JD, ed. Molecular and Clinical Laboratory Immunology. Washington: ASM Press; 2006:1183–92.
49. Choudhary J, Grant SG. Proteomics in postgenomic neuroscience: the end of the beginning. Nat Neurosci 2004;7(5):440–5.
50. Krimbou L, Tremblay M, Davignon J, Cohn JS. Characterization of human plasma apolipoprotein E-containing lipoproteins in the high density lipoprotein size range: focus on pre-beta1-LpE, pre-beta2-LpE, and alpha-LpE. J Lipid Res 1997;38(1):35–48.
51. Ye B, Skates S, Mok SC, et al. Proteomic-based discovery and characterization of glycosylated eosinophil-derived neurotoxin and COOH-terminal osteopontin fragments for ovarian cancer in urine. Clin Cancer Res 2006;12(2):432–41.
52. Paweletz CP, Charboneau L, Bichsel VE, et al. Reverse phase protein microarrays which capture disease progression show activation of pro-survival pathways at the cancer invasion front. Oncogene 2001;20(16):1981–9.
53. EmmertBuck MR, Bonner RF, Smith PD, et al. Laser capture microdissection. Science 1996;274(5289):998–1001.
54. Ekins R, Chu F. Immunoassay and other ligand assays: present status and future trends. J Int Fed Clin Chem 1997;9(3):100–9.
55. Ekins RP, Chu FW. Multianalyte microspot immunoassay–microanalytical "compact disk" of the future. Clin Chem 1991;37(11):1955–67.
56. Wiese R. Analysis of several fluorescent detector molecules for protein microarray use. Luminescence 2003;18(1):25–30.
57. Tannapfel A, Anhalt K, Hausermann P, et al. Identification of novel proteins associated with hepatocellular carcinomas using protein microarrays. J Pathol 2003;201(2):238–49.
58. Sheehan KM, Calvert VS, Kay EW, et al. Use of reverse phase protein microarrays and reference standard development for molecular network analysis of metastatic ovarian carcinoma. Mol Cell Proteomics 2005;4(4):346–55.
59. Yip TT, Hutchens TW. Immobilized metal ion affinity chromatography. Mol Biotechnol 1994;1(2):151–64.
60. Mehta AI, Ross S, Lowenthal MS, et al. Biomarker amplification by serum carrier protein binding. Dis Markers 2003;19(1):1–10.
61. Li JN, White N, Zhang Z, et al. Detection of prostate cancer using serum proteomics pattern in a histologically confirmed population J Urol 2004;171(5):1782–7, Erratum in 172(1):389.
62. Chen R, Pan S, Brentnall TA, Aebersold R. Proteomic profiling of pancreatic cancer for biomarker discovery. Mol Cell Proteomics 2005;4(4):523–33.

63. Rogers MA, Clarke P, Noble J, et al. Proteomic profiling of urinary proteins in renal cancer by surface enhanced laser desorption ionization and neural-network analysis: identification of key issues affecting potential clinical utility. Cancer Res 2003;63(20):6971–83.
64. Sorace JM, Zhan M. A data review and re-assessment of ovarian cancer serum proteomic profiling. BMC Bioinformatics 2003;4:24.
65. Wong YF, Cheung TH, Lo KW, Wang VW, Chan CS, NG TB, Chung TK, Mok SC. "Protein profiling of cervical cancer by protein bio chips: proteomic scoring to discriminate cervical cancer from normal cervix." Cancer Lett 2004: 211(2):227–34.
66. Becker S, Cazares LH, Watson P, et al. Surfaced-enhanced laser desorption/ionization time-of-flight (SELDI-TOF) differentiation of serum protein profiles of BRCA-1 and sporadic breast cancer. Ann Surg Oncol 2004;11(10): 907–14.
67. Gericke B, Raila J, Sehouli J, et al. Microheterogeneity of transthyretin in serum and ascitic fluid of ovarian cancer patients. BMC Cancer 2005;5:133.
68. Ord JJ, Streeter E, Jones A, et al. Phase I trial of intravesical Suramin in recurrent superficial transitional cell bladder carcinoma. Br J Cancer 2005;92(12): 2140–7.
69. Liotta LA, Lowenthal M, Mehta A, et al. Importance of communication between producers and consumers of publicly available experimental data. J Natl Cancer Inst 2005;97(4):310–4.
70. Ransohoff DF. Lessons from controversy: ovarian cancer screening and serum proteomics. J Natl Cancer Inst 2005;97(4):315–9.
71. McShane LM, Altman DG, Sauerbrei W, Taube SE, Gion M, Clark GM. Reporting recommendations for tumor marker prognostic studies. J Clin Oncol 2005;23(36):9067–72.
72. Karsan A, Eigl BJ, Flibotte S, et al. Analytical and preanalytical biases in serum proteomic pattern analysis for breast cancer diagnosis. Clin Chem 2005;51(8):1525–8.
73. Ullrich B, Ushkaryov YA, Sudhof TC. Cartography of neurexins - more than 1000 isoforms generated by alternative splicing and expressed in distinct subsets of neurons. Neuron 1995;14(3):497–507.
74. Alessandro R, Belluco C, Kohn EC. Proteomic approaches in colon cancer: promising tools for new cancer markers and drug target discovery. Clin Colorectal Cancer 2005;4(6):396–402.
75. Petricoin EF, Ardekani AM, Hitt BA, et al. Use of proteomic patterns in serum to identify ovarian cancer. Lancet 2002;359(9306):572–7.
76. Tan AR, Yang X, Hewitt SM, et al. Evaluation of biologic end points and pharmacokinetics in patients with metastatic breast cancer after treatment with erlotinib, an epidermal growth factor receptor tyrosine kinase inhibitor. J Clin Oncol 2004;22(15):3080–90.
77. Posadas EM, Liel MS, Kwitkouski V, Minasian L, Godwin AK, Hussain MM, Espina V, Wood BJ, Steinberg SM, Konn EC. "A phase II study of gefitinib in patients with refractory or recurrent epithelial ovarian cancer." Cancer 2007;109(7):1323–30.
78. Posadas EM, Kwitkouski Kotz HL, Espina V, Minasian L, Tchabo N, Premkumar A, Hussain MM, Chang R, Stemberg SM, Konn EC. "A prospective analysis of imatinib induced c-kit modulation in ovarian cancer: a phase II clinical study with proteomic profiling." Cancer 207;110(2):309–17.

79. Ito M, Harada T, Tanikawa M, Fujii A, Shiota G, Terakawa N. Hepatocyte growth factor and stem cell factor involvement in paracrine interplays of theca and granulosa cells in the human ovary. Fertil Steril 2001;75(5):973–9.
80. Parrott JA, Kim G, Skinner MK. Expression and action of kit ligand/stem cell factor in normal human and bovine ovarian surface epithelium and ovarian cancer. Biol Reprod 2000;62(6):1600–9.

6 Insight on Renal Cell Carcinoma Proteome

Cecilia Sarto, Vanessa Proserpio, Fulvio Magni, and Paolo Mocarelli

Contents

1 What is Renal Cell Carcinoma
2 Genomic Contribution
3 Why Proteomics
4 What is Now Known About the RCC Proteome
5 Advances in Biomarker Discovery
6 Future Implications

Summary

Several efforts are today focused on studying the most wide form of tumor affecting human kidney, renal cell carcinoma (RCC), because of our inability to diagnose and treat this very aggressive neoplasia. Different complementary approaches based on genomic and proteomic tools are used to highlight its altered molecular processes, and new developed methods and techniques are implemented in the search of possible biomarkers. However, notwithstanding the great work done by several groups and the enormous amount of information present in literature, knowledge about its pathogenesis is still incomplete, and several markers of RCC are proposed but not yet validated.

From: *Cancer Drug Discovery and Development
Cancer Proteomics: From Bench to Bedside*
Edited by: S. S. Daoud © Humana Press Inc., Totowa, NJ

Key Words: Renal cell carcinoma; proteome; genomic; biomarkers; two-dimensional electrophoresis; serological proteome analysis; post-translational modification

1. WHAT IS RENAL CELL CARCINOMA

Renal cell carcinoma (RCC) is the most common renal cancer affecting the kidney that accounts for 3% of all human adult tumors and represents about 90% of renal cancer *(1,2)*.

A range of biological and clinical behaviors characterizes RCC. Owing to the absence of early clinical symptoms, kidney cancer is often diagnosed in advanced stages in patients with flank pain, hematuria, or a palpable abdominal mass. Approximately 25% of these patients have metastatic disease. At present, incidental tumor discovery during investigations for other disorders by diagnostic imaging techniques increases RCC diagnosis at the early stages *(2,3)*. The onset of most RCC is sporadic, while the inherited form is less frequent.

Traditionally, radical nephrectomy is the treatment of choice for localized RCC. In the last years, a laparoscopic approach has been introduced for early stage disease, with evident benefits for patients in the postoperative period. Partial nephrectomy or nephron-sparing surgery is indicated in selected cases, including solitary kidney, bilateral RCC, renal failure, and hereditary forms of RCC *(4)*. Although metastatic RCC has a poor prognosis, also due to chemotherapy and radiotherapy resistance, radical nephrectomy and surgical excision of isolated metastasis are also recommended *(5)*. The presence of spontaneous regressions of RCC suggests the possibility to stimulate the immune system through cytokines. Currently, immunotherapy with interleukin 2 (IL-2) and interferon α is respectively approved in United States and in Europe for treatment in advanced RCC, but this therapeutic approach provides positive outcomes only in a minority of patients, due to dissimilar responses related to different tumor types *(6)*.

RCC is a heterogeneous group of cancers, arising from epithelial cells of renal tubules, classified into specific carcinoma cell subtypes: clear cell (70% of all cases), papillary (10–15%), chromophobe (5%), and collecting duct (1%) *(7)*. In addition, mixed subtypes in the same tumor are relatively frequent.

Conventional prognostic factors such as stage and grade, with sub-classifications according to the tumor location, lymph node involvement, and presence of metastases, are used in patient management. The most important prognostic factor for RCC is the tumor, node, and metastasis (TNM) staging system, modified in 1997 by the Union International Contre le Cancer (UICC) and the American Joint Committee on Cancer (AJCC) *(8)*.

The variety and heterogeneity concerning histological subtypes, chromosomal aberrations, gene mutations, gene transcription, and gene products is the cause of the enormous variability of the antigenic pattern of individual RCC and absence of molecular markers for early detection, prognostic information, and treatment guides. Several substances connected with cell cycle, apoptosis, cell regulation, and cell–cell interaction have been evaluated as RCC markers. These are ferritin, nuclear matrix protein 22, neopterin, ki-67, p21, cyclin D, tissue polypeptide antigen, bcl-2, erythropoietin, and fibrinogen, but unfortunately, none of them is qualified as an ideal marker *(9)*.

The need to improve patient outcomes with renal cancer have shifted attention toward gaining more knowledge in the study of renal cancer with new strategies, which employ and combine different genomic, transcriptomic, and proteomic tools to achieve biomarkers for clinical management of RCC.

2. GENOMIC CONTRIBUTION

In the last 10 years, several studies focused on the identification of cancer specific karyotypes and gene alterations for tumors that originate within the kidney *(10)*. The contribution of genetic data with phenotypic and histological observations revealed several distinctive subtypes *(7,11)*. Therefore, a reliable classification based on correlation between chromosomal alterations and histological subtypes has been proposed: (1) deletion of chromosome 3p or addition of chromosome 5q combined with deletion of two or more of chromosomes 6q, 8p, 9p, or 14q for clear-cell RCCs *(12)*; (2) a gain of two or more of chromosomes 3q, 7, 8, 12, 16, 17, or 20 and no 3p loss for papillary RCCs; (3) loss of two or more of chromosomes 1, 2, 6, 10, 13, or 17 for chromophobe RCCs *(13,14)*. The widespread use of DNA microarray technology has generated a large amount of data from various cells and tissues providing the tissue–gene relationship by identification of genes in given tissues. This relationship has been also studied in the screening of different RCC subtypes with diagnostic aims. Recently, one of these studies has shown that inactivation of the tumor suppressor von Hippel–Lindau (VHL) located on chromosome 3p25 is associated with sporadic RCC *(15)*. VHL protein forms a complex with other proteins such as elongin C and B and cul-2 protein (Fig. 1). This tumor suppressor

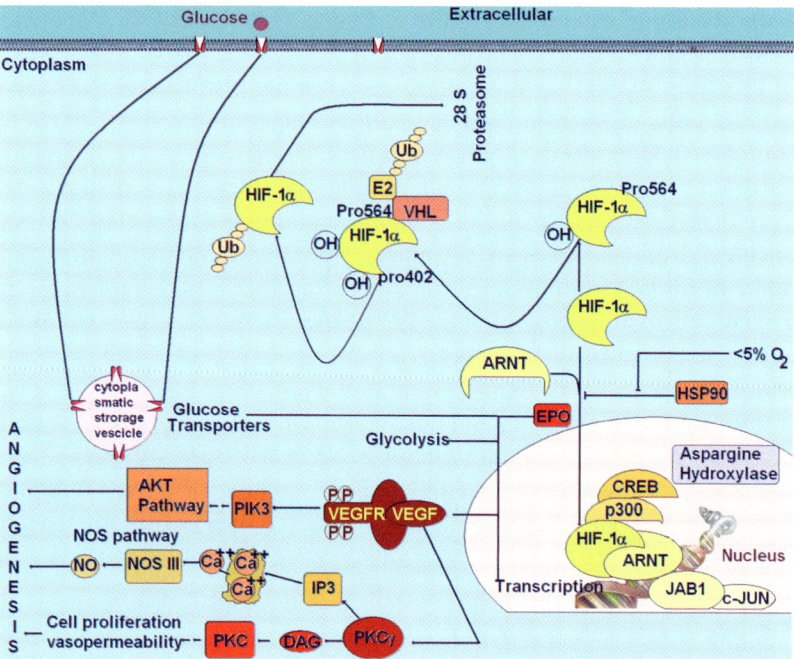

Fig. 1. Illustration of the von Hippel–Lindau pathway.

gene down-regulates the alpha subunits of the hypoxia inducible factors (HIF-1α and HIF-2α), which promote several gene expressions [glucose transport, vascular endothelial growth factor (VEGF), epidermal growth factor, and transforming growth factor-alpha (TGF-α)] *(16)*. The cDNA microarray analysis of more than 7000 genes has shown a down-regulation of mitochondrial and distal nephron genes and overexpression of vimentin, class II major histocompatibility complex (MHC)-related molecules in conventional RCC, while paravalbumin and galectin-3 are found up-regulated in chromophobe RCC/oncocytoma tissues *(17)*. Beginning from 2001, the number of genes studied has increased from 60,000 to 100,000 using Affimetrix microarray techniques *(17–19)*. About 170 genes belonging to the families of cell adhesion, signal transduction, and nucleotide metabolism have been found up-regulated in RCC, whilst 150 genes playing a role in ion homeostasis, oxygen and radical metabolism, and small molecule transport have been observed to be down-regulated. The list of new genes with altered expression discovered using genomic array approaches is continuously increasing, and some of those already identified with modified expression are confirmed by recent publications *(20–24)*. An interesting study shows an overexpression of 230 genes in clear cell RCC including those for VEGF, glucose transporter 1 and 3 (SLC2A1 and SLC2A3), endothelin 1, and insulin-like growth factor binding protein, which are all strictly regulated by HIF-1α as already mentioned. Genes overexpressed in papillary RCC include alpha-methylacyl-CoA racemase, the two oncogenes GRO1 and GRO2, while in chromophobe and oncocytoma they include nicotinamide nucleotide transhydrogenase, fumarate hydratase, solute carrier family 25 members and genes encoding for mitochondrial proteins. Overexpressed are also genes involved in cell adhesion and transport such as laminin A3, fibrinonectin 1, fibrinonectin receptor alpha subunit, vWF, and BIGH3 *(25)* and particularly in aggressive clear cell RCCs the genes IL-8 and survivin *(26)*. Hypermethylation has been evaluated as one of causes of gene alteration, and evidences of the diagnostic potential of DNA methylation-based markers are growing *(27–29)*. Moreover, a significant association between promoter methylation and pathological stage in renal tumors has been shown *(30)*.

Moreover, this relationship has been also studied in RCC subtypes with prognostic value by tissue array constructed for Ki67, p53, gelsolin, carbonic anhydrase (CA)9, CA12, PTEN (phosphatase and tensin homolog deleted on chromosome 10), epithelial cell adhesion molecule, vimentin, S-100, and cyclin B1 *(31–35)*. Current efforts are to integrate molecular information from tissue microarrays with genomic data to generate a molecular integrated staging system.

3. WHY PROTEOMICS

Even so, to date, the ability to predict the outcome of RCC after surgical or systemic therapy is limited and needs to implement new strategies, making the most of proteomic tools and combining genomic, transcriptomic, and proteomic data. Currently, the old but open question "what is a gene?" *(36)* has emphasized the great weight of proteomic studies even if complete identification of DNA coding sequences is now available. Therefore, the proteomic aspects of RCC must be studied first because protein alterations of the living system homeostasis can be better visualized at the proteome level. Additionally, recent studies have demonstrated a lack of correlation between the transcriptional profiles and the actual protein levels in cells *(37)*. As a consequence, abundant protein alterations cannot be predicted based on mRNA levels enforcing the attention to proteome. Moreover, final structure of normal or modified proteins cannot be predicted by the DNA or RNA sequence. Alternative splicing, proteolytic processing, post-translational modification (PTM), trafficking, and protein interaction can modify the chemical structure and/or the biological functions. A valid example is the chaperone Hsp27 found in RCC significantly different in expression and isoform numbers with a 4.5–5.9 pI range and 18–29 kDa Mr range *(38)*. The large number of isoforms in RCC due to the phosphorylation of Hsp27 has been studied by two-dimensional gel electrophoresis (2-DE) Western blot and subsequently by mass spectrometry (MS) analysis. The MS analysis performed by low picomole amounts of phosphopeptides using home-made nanoLC column and ESI-MS/MS analysis has confirmed three phosphorylated protein species of the Hsp27 never directly identified before *(39)*. As the phosphorylation is known to be associated with a decrease of the aggregation state, this suggests modified modulation of Hsp27 functions in the cancer cells. Increased phosphorylation has been demonstrated also for proteins such as moesin and radixin, but using phosphospecific antibodies not directly by MS analysis *(40)*. The phosphorylation of these three proteins plays an important role in the regulation of actin organization. Therefore, evaluation of PTM of proteins within cells or tissues and their variation of content by proteomic approaches can represent a key to identify new diagnostic biomarkers or targets for therapeutic purposes. Moreover, these alterations occurring inside the cells are expected to be reflected in biological fluids (blood and urine) where it is supposed that they may be detected by proteomic techniques. The several proteomic techniques based on the search of differences in tissues, cells, or fluids can be combined with MS: pre-fractionation by depletion of the most abundant proteins; separation by surface-enhanced laser desorption/ionization mass spectrometry (SELDI MS), OFF-GEL, Free Flow, ClinProt technique; labelling by biotin reagents (ICAT) or by fluorophores can be employed before 2-DE *(41,42)*; or fractionation by liquid chromatography.

4. WHAT IS NOW KNOWN ABOUT THE RCC PROTEOME

Various groups are working in the RCC proteome field, and at present, a number of tumor-correlated proteins, mostly downexpressed, have been identified by protein profile comparison between cancer and non-malignant biological samples. Traditionally, proteomics starts with cataloging and developing lists of the cellular protein repertoire of a given tissue that are the base line for the evaluation of differences *(43,44)*. Diverse strategies have been applied for the analysis of renal tissues, primary cell cultures, and culture cell lines, with results that are just beginning. This research involves also biological fluids such as plasma and urine, where potential biomarkers may be quantitatively detected. From the analysis of tissues and cell lines, the majority of identified proteins are housekeeping belonging to the various intracellular compartments such as cytosol, mitochondria, and nucleus *(38,45–49)*. These proteins may be chiefly grouped in three functional classes: (1) enzymes involved in glycolysis and tricarboxylic acid cycles; (2) enzymes that play a role in the mitochondrial respiratory chain; (3) proteins that protect cells from oxidative radicals, in agreement to some data previously reported in the literature. Among the mitochondrial proteins, manganese superoxide dismutase, as well as aconitate hydratase, and proteins involved in the propanate metabolism are more abundant in RCC (Fig. 2). Proteins involved in hypoxia pathway correlated to VHL gene mutations that play an essential role in angiogenesis, glycolysis, and apoptosis show modified expression in RCC, including HIF-1α, CA IX, VEGF, and other important members of the hypoxia-induced gene family such as the septin family members particularly SEPT2, moesin, and fascin that are up-regulated (Fig. 1) *(40)*. Increased levels of Hsp27, lactate dehydrogenase, aldolase A and C, pyruvate kinase M2, thymidine phosphorylase, vimentin, and members of annexin family have been also specifically correlated with RCC, and a number of these alterations have been confirmed by immunohistochemistry *(38,40,50)*.

The search of RCC biomarkers involves also studies to target specific tumor antigens presented by surface MHC class II molecules to prepare a specific vaccine-induced T-cell response *(51)*. One strategy to discovering tumor antigens employs in vivo sensitized tumor-reactive T cells from individual cancer patients, whilst another approach is based on epitope prediction by gene product profiles. The conditions required for defining T-cell-recognized tumor antigens, such as stable T-cell lines and established tumor cell lines, are very difficult to satisfy. Therefore, new methodologies termed SEREX (serological screening of recombinant cDNA expression libraries) *(52)* and SERPA (serological proteome analysis) have been otherwise applied to analyze B-cell response to already known tumor antigens *(53–55)*. SERPA is a proteomic approach that allows to determine the global

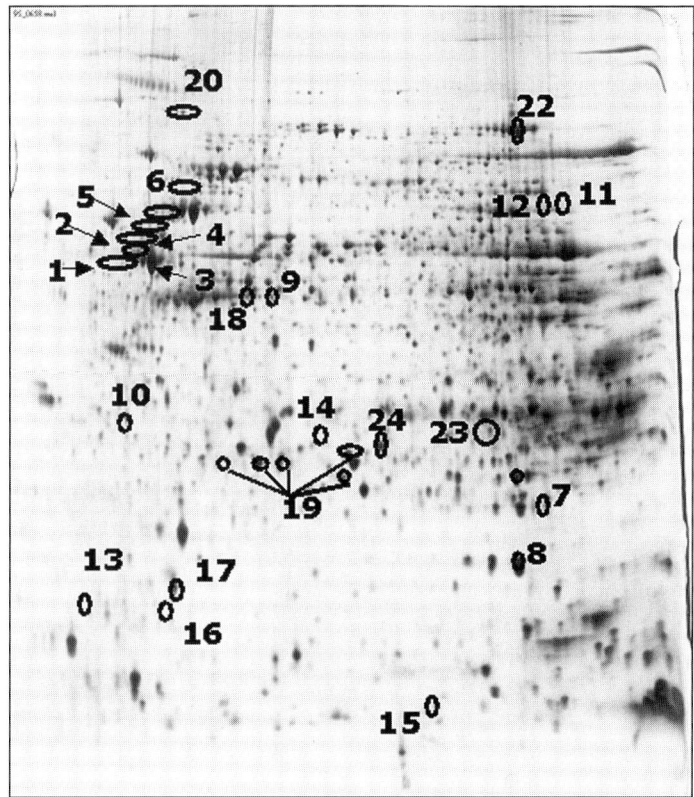

Fig. 2. Image of silver-stained two-dimensional gel electrophoresis gel obtained from renal cell carcinoma tissue showing some of the modified proteins: 1–6, vimentin; 7, manganese superoxide dismutase; 8, alpha crystalline B chain; 9–10 and 16–17, actin fragments; 11–12, fibrinogen beta chain precursor; 13, cytochrome b5 isofrom2; 14, enoil CoA hydratase; 15, beta 2 microglobulin precursor; 18, lamin A/C; 19, Hsp27; 20, major vault protein; 21, triosephosphate isomerse; 22, aconitate hydratase; and 23–24, thioredoxin.

immunoreactivity of patient sera toward tumor proteins separated by 2-DE from tumor tissues. This method has highlighted a relatively low number of tumor-associated antigens in RCC such as aldose reductase, superoxide dismutase 1, thioredoxin, enoyl-CoA hydratase, members of the cytoskeletal family (56), major vault protein, thymidine phosphorylase, annexins I and IV, manganese superoxide dismutase (50) and transgelin, and other six proteins involved in cell contractile function (54,57). In the last months, by applying a differential MS technique for immuno-proteomic analysis of MHC class I ligands in RCC peptide mixtures, Flad and colleagues *(58)* have identified two peptides arising from heme oxygenase-1, a protein involved in apoptosis resistance, immunosuppression, and angiogenesis. In the differential MS strategy, pools of pre-fractionated MHC class I peptides from normal kidney and RCC have been labeled respectively with light and heavy nicotinic acid reagent reacting with the N-terminus, epsilon amino groups of lysine and phenolic groups of tyrosine, and finally, a relative quantification of labeled peptides has been obtained. This method could be a molecular key to identify tumor-specific antigens arising from splicing before their MHC presentation including the PTM not predictable by mRNA analysis *(59)*.

Although after about a decade several tumor antigens have been recovered by these proteomic methodologies, a common specific antigen for RCC has not yet been discovered; only about 70% of patient sera has been found to be responsive against one or two antigens. However, the limits of these methods are in relation with the heterogeneity of RCC. Infact, antibodies against tumor antigens are demonstrated not to be associated with histological type of RCC, and each of them is common only to a restricted number of RCC patients.

5. ADVANCES IN BIOMARKER DISCOVERY

The biomarker story is still in the discovery era, therefore, data have to be improved in light of new technologies and strategies to overcome several difficulties. The major pitfall is the heterogeneity of kidney tissue containing several cell types, and tumor tissue with cells showing a miscellanea of chromosomal aberrations, gene modifications, gene expression, and gene products. Therefore, contamination from surrounding cells present in the renal tissue can conceal marker detection in RCC epithelial cells. To improve the homogeneity and consistency of normal and RCC tissue samples, magnetic microbeads coated with specific epithelial cell antibodies have been utilized and the number of epithelial cells increased *(60)*. Successively, results show about 30 proteins differentially expressed in RCC. Most of these are downexpressed in agreement with the chromosomal aberrations occurring in the neoplastic cells confirming the known data. Moreover, 11 of the identified proteins have mitochondrial location, and 6 are surface and secreted proteins. The differential expression of mitochondrial proteins is in agreement with mitochondrial morphological changes pointed out in RCC by histopathology. Of the overexpressed proteins, annexin A1 (ANXA1) is now evaluated by Western blot with specific antibodies to confirm overexpression in a large number of RCC cases. The ANXA1 is a calcium/phospholipid-binding protein which promotes membrane fusion and is involved in exocytosis and regulates phospholipase A2 activity.

To overcome variability in conventional 2-DE due to running conditions and stain detection, we have introduced 2-D DIGE using two diverse fluorochromes (Cy2, Cy3), two diverse populations, and an internal standard by a third Cy (Cy5) in the same gel. After differential analysis by DeCyder dedicated software, 12/53 spots are found common to those pointed out by conventional 2-DE. These proteins are now being identified by MS and specific antibodies.

Several different analytical approaches have been developed in order to overcome some of the well-known limitations of the classical proteomics strategies. Among these approaches, 2-D DIGE can be applied to improve detection of surface proteins *(61,62)*. The samples can also be enriched in surface proteins through labeling with biotinylated reagents and separation by sepharose-streptavidin affinity column, *(63–67)*. Another old and still most valid method can be subcellular fractionation of membrane proteins that permit to reduce sample complexity and enrich plasma membrane fractions in hydrophobic proteins *(68)*. New 2-DE techniques have been developed in which the first dimension, conventionally based on an immobilized pH gradient isoelectrofocusing, has been substituted by alternative separations to resolve highly hydrophobic proteins including bidimensional blue native/sodium dodecyl sulfate–polyacrylamide gel electrophoresis (SDS–PAGE), double SDS–PAGE, and two-dimensional BAC/SDS electrophoresis *(69–71)*. At present, the application of these new methods has not yet produced useful results for RCC.

Another strategy employs primary cell cultures *(40,72)*. The in vitro culture method offers more homogeneous and enriched cytological material, ability to perform multiple and repeated experiments over long intervals, and the possibility to study metabolic events in viable cells. Examination of specific proteins, SODM, Hsp 27, and results of proteomic and transcriptomic studies provided by real-time PCR gene expression analysis, 2-DE and MS, indicate that primary cell cultures retain, at the first passages, the proteomic differential profile of corresponding tissues.

An attractive alternative in identifying biomarkers for RCC is focused on detection of molecules in different sources from tissues such as serum or plasma *(73)* and urine that are collected noninvasively in large amounts. Different methods and technologies have been employed to separate and identify proteins from urine: 2-DE *(74)*; liquid chromatography applied directly for urine *(75)* or for isolated urinary vesicles, the exosomes *(76)*; and SELDI or ClinProt analysis combined to nano-LC/ESI-MS/MS. To date, few results concerning identification of specific biomarkers have been obtained by analysis of the urine from RCC patients. Cannon et al. *(77)* have evidenced that using high-throughput fluorescent microplate activity assay the measure of elevated levels of matrix metalloproteinases by degradation of collagen IV is not useful for diagnosis of RCC in contrast with previous results, because most of the normal urine samples had the ability to degrade collagen IV. It is true that urine is an easily collected sample, but the time of collection, the storage, and handling are critical in relation with protein degradation and result reproducibility. The application of standardized protocols in the pre-analytical phase should facilitate the discovery of biomarkers in urine by the new technologies.

6. FUTURE IMPLICATIONS

The task and dream is to use the fruits of results obtained by proteomic strategies to create diagnostic platforms readily applicable in clinical management. The means used to derive from proteomic results a set of candidate biomarkers can be estimated in terms of selectivity and specificity by the traditional ELISA test for screening. Connections between the several markers have to be carefully observed and understood by bioinformatics that allow to analyze a huge amount of gene and protein data across the functional pathways in a short time.

The early detection of tumor for timely therapeutic intervention is always a relevant problem in clinical management. The molecular modifications characterizing early stages are still a mystery, and therefore, large-scale screening is impossible. The contribution of proteomic analysis can favor the comprehension of the mechanisms involved in tumorigenesis. It is now known that multiple gene mutations are at the basis of carcinogenesis and determine protein alterations. As the mRNA levels do not always correlate with protein expression, proteomic profiles become crucial for unraveling the behavior of these proteins and of their pathways. An important aid in better understanding these mechanisms is offered by housekeeping proteins that are often found to be differentially expressed in cancer.

Moreover, the biomarkers applied by clinicians to control tumor progression during follow-up do not have a sensitivity and specificity useful for wide application. The numerous proteomic and genomic studies on RCC have produced a great amount of dispersive and not well-correlated data making the interpretation difficult. Furthermore, only a small number of cases have been observed thus far, and the potential markers have not been identified in all, highlighting the high biological variability. To overcome this, it is necessary to subdivide biological markers into clusters which identify specific subgroups of patients. New possibilities toward this goal are offered by platform technologies such as chip technology, SELDI, and so on.

Another obstacle in the search for new RCC biomarkers is represented by the complexity of the tumoral microenvironment. However, the identification of autoantibodies directed toward tumoral antigens has allowed to improve the detection of markers in low concentration and to widen the knowledge concerning the specific immune response to the tumor. The identification of these antigens has favored the development of anticancer vaccines obtained exchanging whole tumoral cells with specific antigens. The clinical response of RCC patients to this treatment that utilizes dendritic cells conjugated to tumoral antigens seems to be encouraging and non-toxic. Furthermore, proteomic tools can provide a valid help in studying the effects of drug treatments to characterize the mechanisms involved in toxic reactions and to identify sensible targets. Additionally, the study of PTM in the diverse phases of the disease, or due to a specific treatment, can be the starting point for specific individual therapeutic strategies.

REFERENCES

1. Jemal A, Siegel R, Ward E, Murray T, Xu J, Smigal C, Thun MJ. Cancer statistics, 2006. CA Cancer J Clin 2006; 56:106–130.
2. Drucker BJ. Renal cell carcinoma: Current status and future prospects. Cancer Treat Rev 2005; 31:536–545.
3. Cohen HT, McGovern FJ. Renal cell carcinoma. N Engl J Med 2005; 353: 2477–2490.
4. Novick A.C. Laparoscopic and partial nephrectomy. Clin Cancer Res 2004; 10:6322S–6327S.
5. Sengupta S, Leibovich BC, Blute ML, Zincke H. Surgery for metastatic renal cell cancer. World J Urol 2005; 23:155–160.
6. Fishman M, Seigne J. Immunotherapy of metastatic renal cell cancer. Cancer Control 2002; 9:293–304.
7. Storkel S, Eble JN, Adlakha K, Amin M, Blute ML, Bostwick DG, Darson M, Delahunt B, Iczkowski K. Classification of renal cell carcinoma: Workgroup No. 1. Union International Contre le Cancer (UICC) and the American Joint Committee on Cancer (AJCC). Cancer 1997; 80:987–989.
8. Srigley JR, Hutter RVP, Gelb AB, Henson DE, Kennedy G, King BF, Raziuddin S, Pisansky TM. Current prognostic factors - Renal cell carcinoma: Workgroup No. 4.

Union International Contre le Cancer (UICC) and the American Joint Committee on Cancer (AJCC). Cancer 1997; 80:994–996.
9. Kashyap MK, Kumar A, Emelianenko N, Kashyap A, Kaushik R, Huang R, Khullar M, Sharma SK, Singh SK, Bhargaave AK, Upadhyaya SK. Biochemical and molecular markers in renal cell carcinoma: An update and future prospects. Biomarkers 2005; 10:258–294.
10. Zbar B, Klausner R, Linehan WM. Studying cancer families to identify kidney cancer genes. Annu Rev Med 2003; 54:217–233.
11. Kovacs G, Akhtar M, Beckwith BJ, Bugert P, Cooper CS, Delahunt B, Eble JN, Fleming S, Ljungberg B, Medeiros LJ, Moch H, Reuter VE, Ritz E, Roos G, Schmidt D, Srigley JR., Storkel S., van den Berg E., Zbar B. The Heidelberg classification of renal cell tumours. J Pathol 1997; 183:131–133.
12. Zhao WP, Gnarra JR, Liu S, Knutsen T, Linehan WM, Whang-Peng J. Renal cell carcinoma. Cytogenetic analysis of tumors and cell lines. Cancer Genet Cytogenet 1995; 82:128–139.
13. Burgert P, Kovacs G. Molecular differential diagnosis of renal cell carcinomas by microsatellite analysis. Am J Pathol 1996; 149:2081–2088.
14. Kovacs G. Molecular differential pathology of renal cell tumours. Histopathology 1993; 22:1–8.
15. Linehan WM, Walther MM, Zbar B. The genetic basis of cancer of the kidney. Urology 2003; 170:2163–2172.
16. Linehan WM, Grubb RL, Coleman JA, Zbar B, Walther MM. The genetic basis of cancer of kidney cancer: Implications for gene-specific clinical management. Br J Urol 2005; 95 (suppl 2):2–7.
17. Young AN, Amin MB, Moreno CS, Lim SD, Cohen C, Petros JA, Marshall FF, Neish AS. Expression profiling of renal epithelial neoplasms: A method for tumor classification and discovery of diagnostic molecular markers. Am J Pathol 2001; 158:1639–1651.
18. Skubitz KM, Skubitz AP. Differential gene expression in renal-cell cancer. J Lab Clin Med 2002, 140:52–64.
19. Boer JM, Huber WK, Sultmann H, Wilmer F, von Heydebreck A, Haas S, Korn B, Gunawan B, Vente A, Fuzesi L, Vingron M, Poustka A. Identification and classification of differentially expressed genes in renal cell carcinoma by expression profiling on a global human 31,500-element cDNA array. Genome Res 2001; 11:1861–1870.
20. Zhao H, Ljungberg B, Grankvist K, Rasmuson T, Tibshirani R, Brooks JD. Gene expression profiling predicts survival in conventional renal cell carcinoma. PLoS Med 2006; 3:e13.
21. Yao M, Tabuchi H, Nagashima Y, Baba M, Nakaigawa N, Ishiguro H, Hamada K, Inayama Y, Kishida T, Hattori K, Yamada-Okabe H, Kubota Y. Gene expression analysis of renal carcinoma: Adipose differentiation-related protein as a potential diagnostic and prognostic biomarker for clear-cell renal carcinoma. J Pathol 2005; 205:377–387.
22. Jones J, Otu H, Spentzos D, Kolia S, Inan M, Beecken WD, Fellbaum C, Gu X, Joseph M, Pantuck AJ, Jonas D, Libermann TA. Gene signatures of progression and metastasis in renal cell cancer. Clin Cancer Res 2005; 11:5730–5739.
23. Sultmann H, von Heydebreck A, Huber W, Kuner R, Buness A, Vogt M, Gunawan B, Vingron M, Fuzesi L, Poustka A. Gene expression in kidney cancer is associated with cytogenetic abnormalities, metastasis formation, and patient survival. Clin Cancer Res 2005; 11:646–655.

24. Takahashi M, Yang XJ, Sugimura J, Backdahl J, Tretiakova M, Qian CN, Gray SG, Knapp R, Anema J, Kahnoski R, Nicol D, Vogelzang NJ, Furge KA, Kanayama H, Kagawa S, Teh BT. Molecular subclassification of kidney tumors and the discovery of new diagnostic markers. Oncogene 2003; 22:6810–6818.
25. Liou LS, Shi T, Duan ZH, Sadhukhan P, Der SD, Novick AA, Hissong J, Skacel M, Almasan A, DiDonato JA. Microarray gene expression profiling and analysis in renal cell carcinoma. BMC Urol 2004; 4:9.
26. Kosari F, Parker AS, Kube DM, Lohse CM, Leibovich BC, Blute ML, Cheville JC, Vasmatzis G. Clear cell renal cell carcinoma: Gene expression analyses identify a potential signature for tumor aggressiveness. Clin Cancer Res 2005; 11: 5128–5139.
27. Yamada D, Kikuchi S, Williams YN, Sakurai-Yageta M, Masuda M, Maruyama T, Tomita K, Gutmann DH, Kakizoe T, Kitamura T, Kanai Y, Murakami Y. Promoter hypermethylation of the potential tumor suppressor DAL-1/4.1B gene in renal clear cell carcinoma. Int J Cancer 2006; 118:916–923.
28. Choi YL, Kang SY, Choi JS, Shin YK, Kim SH, Lee SJ, Bae DS, Ahn G. Aberrant hypermethylation of RASSF1A promoter in ovarian borderline tumors and carcinomas. Virchows Arch 2006; 448:331–336.
29. Hoque MO, Begum S, Topaloglu O, Jeronimo C, Mambo E, Westra WH, Califano JA, Sidransky D. Quantitative detection of promoter hypermethylation of multiple genes in the tumor, urine, and serum DNA of patients with renal cancer. Cancer Res 2004; 64:5511–5517.
30. Gonzalgo ML, Yegnasubramanian S, Yan G, Rogers CG, Nicol TL, Nelson WG, Pavlovich CP. Molecular profiling and classification of sporadic renal cell carcinoma by quantitative methylation analysis. Clin Cancer Res 2004; 10:7276–7283.
31. Lin F, Yang W, Betten M, Teh BT, Yang XJ. The French Kidney Cancer Study. Expression of S-100 protein in renal cell neoplasms. Hum Pathol 2006; 37:462–470.
32. Ikuerowo SO, Kuczyk MA, Mengel M, van der Heyde E, Shittu OB, Vaske B, Jonas U, Machtens S, Serth J. Alteration of subcellular and cellular expression patterns of cyclin B1 in renal cell carcinoma is significantly related to clinical progression and survival of patients. Int J Cancer 2006; 119(4): 867–74.
33. Amin MB, Amin MB, Tamboli P, Javidan J, Stricker H, de-Peralta Venturina M, Deshpande A, Menon M. Prognostic impact of histologic subtyping of adult renal epithelial neoplasms: An experience of 405 cases. Am J Surg Pathol 2002; 26: 281–291.
34. Young AN, de Oliveira Salles PG, Lim SD, Cohen C, Petros JA, Marshall FF, Neish AS, Amin MB. Beta Defensin-1, Parvalbumin, and Vimentin. A panel of diagnostic immunohistochemical markers for renal tumors derived from gene expression profiling studies using cDNA microarrays. Am J Surg Pathol 2003; 27:199–205.
35. Li G, Cuilleron M, Gentil-Perret A, Cottier M, Passebosc-Faure K, Lambert C, Genin C, Tostain J. Rapid and sensitive detection of messenger RNA expression for different diagnosis of renal cell carcinoma. Clin Cancer Research 2003; 9: 6441–6446.
36. Paerson H. What is a gene? Nature 2006; 441:399–401.
37. Madi A, Pusztahelyi T, Punyiczki M, Fesus L. The biology of the post-genomic era: The proteomics. Acta Biol Hung 2003; 54:1–14.

38. Sarto C, Valsecchi C, Magni F, Tremolada L, Arizzi C, Cordani N, Cesellato S, Doro G, Favini P, Perego RA, Raimondo F, Ferrero S, Mocarelli P, Galli-Kienle M. Expression of heat shock protein 27 in human renal cell carcinoma. Proteomics 2004; 4:2252–2260.
39. Tremolada L, Magni F, Valsecchi C, Sarto C, Mocarelli P, Perego R, Cordani N, Favini P, Galli Kienle M, Sanchez JC, Hochstrasser DF, Corthals GL. Characterization of heat shock protein 27 phosphorylation sites in renal cell carcinoma. Proteomics 2005; 5:788–795.
40. Craven RA, Stanley AJ, Hanrahan S, Dods J, Uwin R, Totty N, Harnden P, Eardley I, Selby PJ, Banks RE. Proteomic analysis of primary cell lines identifies protein changes present in renal cell carcinoma. Proteomics 2006; 6:2853–2864.
41. Lin YW, Lin CY, Lai HC, Chiou JY, Chang CC, Yu MH, Chu TY. Plasma proteomic pattern as biomarkers for ovarian cancer. Int J Gynecol Cancer 2006; 16 Suppl 1, 139–146.
42. Theodorescu D, Wittke S, Ross MM, Walden M, Conaway M, Just I, Mischak H, Frierson HF. Discovery and validation of new protein biomarkers for urothelial cancer: A prospective analysis. Lancet Oncol 2006; 7:230–240.
43. Sarto C, Marocchi A, Sanchez JC, Giannone D, Frutiger S, Golaz O, Wilkins MR, Doro G, Cappellano F, Hughes G, Hochstrasser DF, Mocarelli P. Renal cell carcinoma and normal kidney protein expression. Electrophoresis 1997; 18: 599–604.
44. Magni F, Sarto C, Valsecchi C, Casellato S, Bogetto SF, Bosari S, Di Fonzo A, Perego RA, Corizzato M, Doro G, Galbusera C, Rocco F, Mocarelli P, Galli Kienle M. Expanding the proteome two-dimensional gel electrophoresis reference map of human renal cortex by peptide mass fingerprinting. Proteomics 2005; 5:816–825.
45. Hwa JS, Kim HJ, Goo BM, Park HJ, Kim CW, Chung KH, Park HC, Chang SH, Kim YW, Kim DR, Cho WS, Kang KR. The expression of ketohexokinase is diminished in human clear cell type of renal cell carcinoma. Proteomics 2006; 6:1077–1084.
46. Seliger B, Lichtenfels R, Atkins D, Bukur J, Halder T, Kersten M, Harder A, Ackermann A, Malenica B, Brenner W, Zobawa M, Lottspeich F. Identification of fatty acid binding proteins as markers associated with the initiation and/or progression of renal cell carcinoma. Proteomics 2005; 5:2631–2640.
47. Hwa JS, Park HJ, Jung JH, Kam SC, Park HC, Kim CW, Kang KR, Hyun JS, Chung KH. Identification of proteins differentially expressed in the conventional renal cell carcinoma by proteomic analysis. J Korean Med Sci 2005; 20:450–455.
48. Sarto C, Deon C, Hochstrasser DF, Mocarelli P, Sanchez JC. Contribution of proteomics to the molecular analysis of renal cell carcinoma with an emphasis on manganese superoxide dismutase. Proteomics 2001; 1:1288–1294.
49. Sarto C, Frutiger S, Cappellano F, Sanchez JC, Doro G, Catanzaro F, Hughes GJ, Hochstrasser DF, Mocarelli P. Modified expression of plasma glutathione peroxidase and manganese superoxide dismutase in human renal cell carcinoma. Electrophoresis 1999, 20:3458–3466.
50. Unwin RD, Hamden P, Pappin D, Rahman D, Whelan P, Craven RA, Selby PJ, Banks RE. Serological and proteomic evaluation of antibody responses in the identification of tumor antigens in renal cell carcinoma. Proteomics 2003; 3:45–55.
51. Watts C. Antigen processing in the endocytic compartment. Curr Opin Immunol 2001; 13:26–31.

52. Sahin U, Türeci O, Schmitt H, Cochlovius B, Johannes T, Schmits R, Stenner F, Luo G, Schobert I, Pfreundschuh M. Human neoplasms elicit multiple specific immune responses in the autologous host. Proc Natl Acad Sci USA 1995; 92:11810–18113.
53. Devitt G, Meyer C, Wiedemann N, Eichmuller S, Kopp-Schneider A, Haferkamp A, Hautmann R, Zoller M. Serological analysis of human renal cell carcinoma. Int J Cancer 2006; 118:2210–2219.
54. Klade CS, Voss T, Krystek E, Ahorn H, Zatloukal K, Pummer K, Adolf GR. Identification of tumor antigens in renal cell carcinoma by serological proteome analysis. Proteomics 2001; 1:890–898.
55. Scanlan MJ, Gordan JD, Williamson B, Stockert E, Bander NH, Jongeneel V, Gure AO, Jäger D, Jäger E, Knuth A, Chen Y-T, Old LJ. Antigens recognized by autohologous antibody in patients with renal-cell carcinoma. Int J Cancer 1999; 83:456–464.
56. Kellner R, Lichtenfelds R, Atkins D, Bukur J, Ackermann A, Beck J, Brenner W, Melchior S, Seliger B. Targeting of tumor associated antigens in renal cell carcinoma using proteome-based analysis and their clinical significance. Proteomics 2002; 2:1743–1751.
57. Lichtenfelds R, Kellner R, Atkins D, Bukur J, Ackermann A, Beck J, Brenner W, Melchior S, Seliger B. Identification of metabolic enzymes in renal cell carcinoma utilizing PROTEOMEX analyses. Biochim Biophys Acta 2003, 1646:21–31.
58. Flad T, Mueller L, Dihazi H, Grigorova V, Bogumil R, Beck A, Thedieck C, Mueller GA, Kalbacher H, Mueller CA. T cell epitope definition by differential mass spectrometry: Identification of a novel, immunogenic HLA-B8 ligand directly from renal cancer tissue. Proteomics 2006; 6:364–374.
59. Hanada K, Yewdell JW, Yang JC. Immune recognition of a human renal cancer antigen through post-translational protein splicing. Nature 2004; 427:252–256.
60. Sarto C, Valsecchi C, Mocarelli P. Renal cell carcinoma: Handling and treatment. Proteomics 2002; 2:1627–1629.
61. Mayrhofer C, Krieger S, Allmaier G, Kerjaschki D. DIGE compatible labelling of surface proteins on vital cells in vitro and in vivo. Proteomics 2006; 6:579–585.
62. Shin BK, Wang H, Yim AM, Le Naour F, Brichory F, Jang JH, Zhao R, Puravs E, Tra J, Michael CW, Misek DE, Hanash SM. Global profiling of the cell surface proteome of cancer cells uncovers an abundance of proteins with chaperone function. Biol Chem 2003; 278:7607 7616.
63. Domon B, Aebersold R. Mass spectrometry and protein analysis. Science 2006; 312:212–217.
64. Giepmans BN, Adams SR, Ellisman MH, Tsien RY. The fluorescent toolbox for assessing protein location and function. Science 2006; 312:217–224.
65. Zhang W, Zhou G, Zhao Y, White MA, Zhao Y. Affinity enrichment of plasma membrane for proteomics analysis. Electrophoresis 2003; 24:2855–2863.
66. Jang JH, Hanash S. Profiling of the cell surface proteome. Proteomics 2003; 3:1947–1954.
67. Chen WN, Yu LR, Strittmatter EF, Thrall BD, Camp DG II, Smith RD. Detection of in situ labeled cell surface proteins by mass spectrometry: Application to the membrane subproteome of human mammary epithelial cells. Proteomics 2003; 3:1647–1651.
68. Raimondo F, Ceppi P, Guidi K, Masserini M, Foletti C, Pitto M. Proteomics of plasma membrane microdomains. Expert Rev Proteomics 2005; 2:793–807.

69. Zahedi RP, Meisinger C, Sickmann A. Two-dimensional benzyldimethyl-n-hexadecylammonium chloride/SDS-PAGE for membrane proteomics. Proteomics 2005; 5:3581–3588.
70. Brouillard F, Bensalem N, Hinzpeter A, Tondelier D, Trudel S, Gruber AD, Ollero M, Edelman A. Blue native/SDS-PAGE analysis reveals reduced expression of the mClCA3 protein in cystic fibrosis knock-out mice. Mol Cell Proteomics 2005; 4:1762–1775.
71. Rais I, Karas M, Schagger H. Two-dimensional electrophoresis for the isolation of integral membrane proteins and mass spectrometric identification. Proteomics 2004; 4:2567–2571.
72. Perego RA, Bianchi C, Corizzato M, Eroini B, Torsello B, Valsecchi C, Di Fonzo A, Cordani N, Favini P, Ferrero S, Pitto M, Sarto C, Magni F, Rocco F, Mocarelli P. Primary cell cultures arising from normal kidney and renal cell carcinoma retain the proteomic profile of corresponding tissues. J Proteome Res 2005; 4:1503–1510.
73. Rogers MA, Clarke P, Noble J, Munro NP, Paul A, Selby PJ, Banks RE. Proteomic profiling of urinary proteins in renal cancer by surface enhanced laser desorption ionization and neural-network analysis: Identification of key issues affecting potential clinical utility. Cancer Res 2003; 63:6971–6983.
74. Thongboonkerd V, Klein JB, Jeans AW, McLeish KR. Urinary proteomics and biomarker discovery for glomerular diseases. Contrib Nephrol 2004; 141:292–307.
75. Soldi M, Sarto C, Valsecchi C, Magni F, Proserpio V, Ticozzi D, Mocarelli P. Proteome profile of human urine with two-dimensional liquid phase fractionation. Proteomics 2005; 5:2641–2647.
76. Pisitkun T, Shen RF, Knepper AA. Identification and proteomic profiling of exosomes in human urine. Proc Natl Acad Sci USA 2004; 101:13368–13373.
77. Cannon GM Jr, Getzenberg RH. Urinary matrix metalloproteinase activity is not significantly altered in patients with renal cell carcinoma. Urology 2006; 67:848–850.

7 Proteomics in Lung Cancer

M. A. Reymond, M. Beshay, and H. Lippert

CONTENTS

1. INTRODUCTION
2. EXPRESSION PROTEOMICS STUDIES
3. FUNCTIONAL PROTEOMICS STUDIES
4. CONCLUSION

SUMMARY

Four main histologic types of lung carcinoma usually are distinguished: squamous cell, undifferentiated small cell, undifferentiated large cell and adenocarcinoma. Proteomics studies are useful complement to histopathology to elucidate the mechanisms that determine clinical phenotype. Several proteomics expression studies have been reported. (1) two-dimensional polyacrylamide gel electrophoresis studies have established the protein profiles of normal, metaplasia, dysplasia and carcinoma tissues of human bronchial epithelia. (2) surface-enhanced laser desorption/ionization-time of flight and gel-free mass spectrometry profiling systems were applied to the discovery of lung cancer biomarkers in serum or plasma. Some innovative proteomics approaches have been proposed in lung cancer cancer. (3) Disease-related volatile organic compounds that might be helpful for the early detection of bronchiogenic cancer were detected in human breath. (4) Proteins were profiled in malignant pleural effusion. Functional studies have highlighted modifications of several kinases and cell cycle proteins in lung cancer, some of them might have therapeutic potential. These studies have also suggested additional cellular processes that were

From: *Cancer Drug Discovery and Development*
Cancer Proteomics: From Bench to Bedside
Edited by: S. S. Daoud © Humana Press Inc., Totowa, NJ

not previously known in lung cancer and have identified tumor-associated antigens. Proteomics techniques have been applied for identifying responders to cisplatinum-based chemotherapy. Protein expression profiles have been shown to predict the outcome of patients with early stage lung cancer. It is expected that proteomics technologies have a great potential for clinical progress in lung cancer.

Key Words: Lung cancer; proteomics; hnRNP; cytokeratins; LCM

1. INTRODUCTION

Lung cancer is the leading cause worldwide of cancer-related death among males and the second among females. Cigarette smoking accounts for over 90% of cases in men and about 70% in women, with a strong dose–response relationship and regression of incidence after quitting. A dose–response relationship with smoking has been shown in the three commonest types of lung cancer: squamous cell, small cell and adenocarcinoma. A small proportion of lung cancers is related to occupational agents, often overlapping with smoking.

Clinical manifestations of lung cancer depend on tumor localization and type of spread. Cough is usually present. In patients with chronic bronchitis, increased intensity and intractability of pre-existing cough suggest a neoplasm. The principal sources of diagnostic information are the patient history and the chest X-ray. X-ray patterns depend on the site of involvement. There is no effective early detection method available, due to lack of screening test. This is unfortunate because about 80% patients in early stage (UICC stage I) could be cured.

Four histologic types of lung carcinoma usually are distinguished (Table 1): (1) squamous cell, frequently arising in the larger bronchi; (2) undifferentiated small cell lung carcinoma (SCLC), producing early hematogeneous metastases; (3) undifferentiated large cell; and (4) adenocarcinoma, usually peripheral. Eighty percent of lung cancer cases are non-small cell lung cancer (NSCLC), and 20% are SCLC. Regardless of subtype, the 5-year survival rate for lung cancer is among the lowest of all cancers at 10–15%.

It is expected that advances in the molecular classification of lung cancer and the identification of causative genetic alterations will lead to improvements in diagnosing and treating patients. Numerous gene expression studies have been published in lung cancer [reviewed in ref. *(1)*], involving numerous translational research technologies.

Table 1
World Health Organization Lung Cancer Classification

I. Epithelial tumors
   A. Benign
      1. Papillomas
      2. Adenomas
   B. Dysplasia/carcinoma in situ
   C. Malignant
      1. Squamous cell carcinoma
         a. Spindle cell variant
      2. Small cell carcinoma
         a. Oat cell carcinoma
         b. Intermediate cell type
         c. Combined oat cell carcinoma
      3. Adenocarcinoma
         a. Acinar
         b. Papillary
         c. Bronchioalveolar
         d. Solid carcinoma with mucin formation
      4. Large cell carcinoma
         a. Giant cell carcinoma
         b. Clear cell carcinoma
      5. Adenosquamous carcinoma
      6. Carcinoid tumor
      7. Bronchial gland carcinoma
      8. Others
II. Soft tissue tumors
III. Mesothelial tumors
   A. Benign
   B. Malignant
IV. Miscellaneous tumors
   A. Benign
   B. Malignant
V. Secondary tumors
VI. Unclassified tumors
VII. Tumor-like lesions

Adapted from ref. *(52)*.

In lung cancer, the relationship between gene expression measured at the mRNA level and the corresponding protein level is complex. When mRNA and protein expression was compared for a cohort of genes in the 76 lung adenocarcinomas and 9 non-neoplastic lung tissues, only 17% of the 165 protein spots and 21.4% of 98 genes had a statistically significant correlation between protein and mRNA expression. Correlation coefficient values were not related to protein abundance. Furthermore, no significant correlation between mRNA and protein expression was found if the average levels of mRNA or protein among all samples were applied across all protein spots or genes. The mRNA/protein correlation coefficient also varied among proteins with multiple isoforms, indicating potentially separate isoform-specific mechanisms for the regulation of protein abundance.

The observations above underscore the need for translational studies combining transcriptomics and proteomics tools. Moreover, transcriptomics technologies do not allow the determination of the tridimensional structure of molecules, so that they cannot give an accurate picture of the functional proteins in the cell. Finaly, post-translational modification of proteins—like phosphorylation or cleavage—is a common feature in normal lung tissue as well as in cancer and cannot be directly detected by genetic studies.

2. EXPRESSION PROTEOMICS STUDIES

The proteome is defined as the total set of proteins expressed by a given cell at a given point of time. Direct separation and quantification of the proteome of human lung cancer cells and paired normal bronchial or alveolar epithelial tissue is a valuable approach for identification of proteins differentially expressed in lung carcinogenesis, for establishing a human lung cancer proteome database and for screening biomarker of disease. As detailed below, expression proteomics studies have proven that substantial heterogeneity in lung cancer is resulting from post-translational modifications. As some findings correlate with patient survival and other clinical parameters, they underscore the critical importance of proteomics studies as a complement for genomic studies in clinical cancer research.

2.1. Squamous Carcinoma

Differential proteome profiles of human lung squamous carcinoma tissue compared to paired tumor-adjacent normal bronchial epithelial tissue were established and analyzed by two-dimensional polyacrylamide gel electrophoresis (2D-PAGE) and matrix-assisted laser desorption/ionization time of flight (MALDI-TOF) mass spectrometry (MS). Forty-three differential proteins were characterized: some proteins were related to oncogenes, and others involved in the regulation of cell cycle and signal transduction *(2,3)*. To get a better insight into the complex multiple-stage process of carcinogenesis of human lung squamous carcinoma, 2D-PAGE profiles of 32 normal, metaplasia, dysplasia and carcinoma tissues of human bronchial epithelia were established. The protein patterns were compared and showed significant differences in the number of average protein spots among the four groups. Twenty-three protein spots selected from dysplasia and invasive carcinoma groups were fingerprinted by MALDI-TOF MS, and database search identified various kinases, kinase inhibitors, metalloproteinase receptors and tumor antigens. Indeed, these proteins might be related to carcinogenesis, and these results provide a fundamental basis for further studies *(4)*.

Another study in early tumor stages and premalignant bronchial lesions has involved the use of surface-enhanced laser desorption/ionization (SELDI) TOF MS. Bronchoscopic biopsies of normal lung, atypical adenomatous hyperplasia and malignant tumors were taken from patients participating in a screening program in Florida. After processing these biopsies with laser capture microdissection (LCM) to obtain pure cell populations, SELDI MS was used to generate protein profiles in each epithelial cell type. Three protein peaks at 17–23 kDa mass from tumor cells were markedly increased compared with normal cells, one of them was not detected in any of the normal cells, but was present at low levels in the atypical cell samples. The authors concluded that these "malignant" protein signatures might allow the identification of populations at high risk for lung cancer *(5)*.

Heterogeneous nuclear ribonucleoprotein (hnRNP) B1 is a RNA-binding protein of 37 kDa and is specifically overexpressed in the nuclei of human lung cancer cells, particularly in squamous cell carcinoma *(6)*. hnRNP is also overexpressed in roentgenographically occult cancers of the lungs and premalignant lesions, such as bronchial dysplasia *(7)*. Overexpression of hnRNP B1 protein was observed in 100% of stage I lung cancer tissues, but it was not found in normal bronchial epithelium. Squamous cell carcinoma of the lungs showed stronger staining than other histological types, and elevation of hnRNP B1 was found in both roentgenographically occult lung cancers and bronchial dysplasia. Furthermore, cytological examination with anti-hnRNP B1 antibody detected cancer cells in sputum, suggesting the potential of hnRNP B1 protein as a new biomarker in the early stage of lung cancer. Because strong staining of hnRNP B1 was also observed in various squamous cell carcinomas of oral and esophageal tissues, overexpression of hnRNP B1 seems to be a common event in the carcinogenic processes of squamous cell carcinoma *(8)*.

2.2. Lung Adenocarcinoma

2D-PAGE and MS were used to identify proteins showing increased expression in lung adenocarcinoma. A series of 93 lung adenocarcinomas (64 UICC stage I and 29 stage III) and 10 uninvolved lung samples were examined for quantitative differences in protein expression using 2D-PAGE and MALDI-TOF MS or peptide sequencing. Nine proteins isoforms found to be overexpressed in the lung tumors were identified as antioxidant enzyme AOE372, ATP synthase subunit d (ATP5D), beta1, 4 galactosyltransferase, cytosolic inorganic pyrophosphatase, glucose-regulated M(r) 58,000 protein, glutathione-S-transferase M4, prolyl 4-hydroxylase beta subunit, triosephosphate isomerase and ubiquitin thiolesterase (UCHL1) *(9)*.

Keratins were the object of a specific proteomics study in lung adenocarcinoma. Cytokeratins (CK) are intermediate filaments whose expression is often altered in epithelial cancer. Systematic identification of lung adenocarcinoma proteins using 2D-PAGE and MS has uncovered numerous CK isoforms. In 93 lung adenocarcinomas (64 stage I and 29 stage III) and 10 uninvolved lung samples, 14 of 21 isoforms of CK 7, 8, 18 and 19 occurred at significantly higher levels in tumors compared to uninvolved adjacent tissue. Specific isoforms of the four types of CK identified correlated with either clinical outcome or individual clinical pathological parameters. All five of the CK7 isoforms associated with patient survival represented cleavage products. Two of five CK7 isoforms, one of eight CK8 isoforms (no. 439) and one of three CK19 isoforms (no. 1955) were associated with survival and significantly correlated to their mRNA levels, suggesting that transcription underlies overexpression of these CK isoforms *(10)*. These data indicate that specific isoforms of individual CK may have utility as diagnostic or predictive markers in lung adenocarcinomas.

2.3. Lung Cancer Cell Lines

Thirty lung cancer cell lines with three different histological backgrounds (squamous cell carcinoma, SCLC and adenocarcinoma) were subjected to two-dimensional difference gel electrophoresis to elucidate the mechanisms that determine histological phenotype. Hierarchical clustering analysis and principal component analysis divided the cell lines according to their original histology. Spot ranking analysis using a support vector machine algorithm and unsupervised classification methods identified 32 protein spots essential for the classification. The proteins corresponding to the spots were identified by MS. Next, lung cancer cells isolated from tumor tissue by laser microdissection (Section 2.4.) were classified on the basis of the expression pattern of these 32 protein spots. Based on the expression profile of the 32 spots, the isolated cancer cells were categorized into three histological groups: the squamous cell carcinoma group, the adenocarcinoma group and a group of carcinomas with other histological types, demonstrating the utility of quantitative proteomic analysis for molecular diagnosis and classification of lung cancer cells *(11)*.

2.4. Sample Preparation

The quality and reproducibility of translational studies depend largely on adequate sample selection and preparation. For example, accurate molecular profiling of a squamous cancer would require comparison with bronchial mucosa obtained in the neighborhood, whereas adenocarcinoma would require lung parenchyma as a control tissue. As lung cancer usually develops in patients suffering chronic bronchitis, obtaining normal control tissue might represent a particular challenge. In any case, sample heterogeneity and in particular tumor tissue heterogeneity needs to be carefully addressed before analysis.

In lung cancer, most proteomics studies available have been performed using whole tissue biopsies. The reliability of results derived from whole-tissue samples depends on the proportion of the various cell populations (epithelial, endothelial, inflammatory cells, etc). Proteomics studies require relatively large samples, so that the risk of analyzing a mixture of normal cells, pathological cells, stromal cells and connective tissue should not be underestimated. Several technical solutions for this problem such as LCM *(12)* and successful purification of epithelial cells using antibody-coated magnetic beads *(13)* have been recently described. However, with one exception *(14)*, they have not been applied in lung cancer proteomics studies so far.

An alternative, novel approach for the evaluation of changes in the lung proteome is the direct evaluation tissue biopsies by imaging MS *(15)*. MALDI-TOF MS was applied directly to 1-mm regions of single frozen tissue section for profiling of protein expression in 79 lung tumors and 14 normal lung tissues, and more than 1600 protein peaks were obtained. A class prediction model was developed with the proteomic patterns in a training cohort of 42 lung tumors and eight normal lung samples. Then, this predictive model was applied to a blinded test cohort, including 37 lung tumors and 6 normal lung samples. It was possible to classify lung cancer histologies, distinguish primary tumors from metastases to the lung from other sites and classify nodal involvement with 85% accuracy in the training cohort. This model nearly perfectly classified samples in the independent blinded test cohort.

2.5. Protein Preparation

After obtaining samples, the cells or tissue contents must be entirely solubilized in order to extract a representative pool of the proteome. In practice, this represents a critical problem: extraction of membrane proteins represents a particular challenge in proteomics research, due to their low solubility (16). In lung cancer, improved deoxycholate-trichloroaetic acid (DOC-TCA) precipitation was used to extract and purify the total proteins of bronchial epithelial samples. When 2D-PAGE was repeated for three times, the average matching rate was 89.3%, an outstanding value, and protein spots in the three gels had a good reproducibility. The average position deviation of matched spots in different gels was low both in the first (isoelectric focusing) and in the second [sodium dodecyl sulfate (SDS)–PAGE] direction (17). Improved DOC-TCA precipitation appears to be so far the only method for protein sample preparation that has been specifically tested in bronchial epithelial tissues.

2.6. Proteomics of Body Fluids

Proteomics technologies were also applied for the analysis of body fluids in lung cancer patients, in particular for the analysis of plasma, serum pleural effusions. Plasma is the most abundant source of proteins in the human body, and is also one of the easiest to collect, leading to its broad use in proteomics research as well as in clinical diagnostics. However, the plasma proteome is also the most difficult version of the human proteome: low abundance biomarkers are obscured by the presence of ubiquitous proteins—the 10 most abundant proteins in plasma account for about 90% of the total proteome. Moreover, the protein content of body fluids is influenced by a large number of different factors, such as protein turnover, dilution, oxydation or degradation and, in pleural effusions, by the influx of plasma proteins.

2.7. Serum

Gradient polyacrylamide gel (SDS–PAGE) was applied to investigate if serum proteins pattern is capable to discriminate lung cancer patients from healthy persons. Serum samples obtained from 66 lung cancer patients and from 44 healthy donors were compared, and different proteins bands were found with increased frequency and/or intensity in both patients and controls. Unfortunately, peptides of interest were not identified *(18)*.

Novel technologies such as SELDI TOF MS ProteinChip system and other protein arrays are now available and have boosted serum protein analysis. In lung cancer, a total of 208 serum samples, including 158 lung cancer patients and 50 healthy individuals, were analyzed by SELDI technology [weak cation exchange (WCX2) chips]. Five protein peaks were automatically chosen as a biomarker pattern in a training set and, when the peptide pattern was tested with the blinded test set, it yielded a sensitivity of 87% and a specificity of 80%. These first results suggested that serum SELDI protein profiling can distinguish lung cancer patients, especially NSCLC patients, from normal subjects with relatively high sensitivity and specificity. However, the identity of the peaks of interest has not been published so far *(19)*. In another study involving 28 serum samples from patients with NSCLC and 12 from normal individuals, two biomarkers were up-regulated while three biomarkers were down-regulated in the serum samples from NSCLC patients *(20)*. This finding is innovative as it would imply that cancer might be detected by the absence of genuine proteins—the opposite of tumor-associated antigens! We would like to note here that we have done similar observations in gastric cancer *(21)*.

Analysis of serum proteins in lung cancer is also possible without gel separation, by coupling for example a two-dimensional microflow liquid chromatography with a tandem MS (2D microLC-MS/MS). A research group incorporated a linear ion-trap mass spectrometer into the microLC-MS/MS system in order to obtain highly improved sensitivity and resolution in MS/MS acquisition and analyzed the proteome of albumin or immunoglobulin-depleted plasma samples from healthy individuals and lung adenocarcinoma patients. Over 100 different proteins were detected, and protein identification datasets of both healthy and adenocarcinoma groups revealed that several proteins could be candidate disease markers *(22)*.

2.8. Pleural Effusion

Pleural effusion, an accumulation of pleural fluid, contains proteins originated from plasma filtrate and, especially when tissues are damaged, parenchyma interstitial spaces of lungs and/or other organs. Because it allows to short coming the problems of overwhelming abundant proteins in serum, the use of proteomics analysis of human pleural fluid for the search of new lung cancer marker proteins, and for their simultaneous display and analysis in patients suffering from lung disorders, is an interesting approach. Two proteomics studies of pleural effusion have been published almost simultaneously.

In the first study, two-dimensional nano-high performance liquid chromatography electrospray ionization tandem MS allowed the identification of 124 proteins, some proteins having not been reported in plasma *(23)*. In the second study, a composite sample was prepared by pooling pleural effusions from seven lung adenocarcinoma patients and analyzed with 2D-PAGE and LC-tandem MS. Forty-four proteins were identified, out of which 7 proteins, retinoblastoma binding protein 7, synaptic vesicle membrane protein, corticosteroid-binding globulin precursor, the positive regulatory (PR)-domain containing protein 11, envelope glycoprotein, MSIP043 protein and titin may be specifically present in pleural effusion *(24)*.

2.9. Bronchoalveolar Lavage Fluid

Bronchoalveolar lavage fluid (BALF) is presently the most common way of sampling the components of the epithelial lining fluid and the most faithful reflect of the protein composition of the pulmonary airways. Proteomics studies of BALF have already been published and contribute both to a better knowledge of the lung structure at the molecular level and to the study of lung disorders at the clinical level *(25)*. 2D-PAGE offers the possibility to simultaneously display and analyze proteins contained in BALF. Protein content of BALF displays a remarkably high complexity, not only due to the wide variety of the proteins it contains but also because of the great diversity of their cellular origins *(26)*.

Early proteome investigations of BALF done to investigate alveolar proteinosis resulted in a 2D-PAGE database of normal BALF published in 1979 *(27)*. Technological advances and development of web databases such as SWISS-2D-PAGE database containing compiled maps of human BALF *(28–30)* boasted advances in the proteomic analysis of BALF. A 2D-PAGE protein database of BALF samples has been developed with a listing of the proteins already identified and of their level and/or post-translational alterations in lung disorders *(31)*. The current master gel of BALF proteins encompasses greater than 1200 spots visualized by silver staining.

2.10. Exhaled Air

BALF is invasive and therefore cannot be routinely employed for probe sampling for screening populations. Based on the hypothesis that aerosol particles excreted in human breath reflect the composition of the BALF, experiments were performed to concentrate and analyze these aerosols directly using a noninvasive technique. As early as 1993, the first 2D-PAGE study was published that analyzed condensates of exhaled air, but MS technology was not available to further develop these first data *(32)*. Results obtained in the meantime suggest that disease-related volatile organic compounds found in human breath could be a very useful diagnostic tool for the early detection of bronchogenic carcinoma. So-called electronic noses are routinely used in commercial applications, including detection and analysis of volatile organic compounds in the food industry. In a discovery study applying electronic noses in lung cancer research, exhaled breath of 14 individuals with bronchiogenic carcinoma and 45 control subjects was analyzed. Principal components and discriminant analysis of the sensor data were used to determine whether exhaled gases could distinguish between cancer and noncancer, and a classification model developed. Then, this predictive model was applied prospectively in a separate group of individuals with and without cancer. In the validation study, the electronic nose had 71% sensitivity and 92% specificity for detecting lung cancer. The results provide feasibility to the concept of using the electronic nose for managing and detecting lung cancer *(33)*.

A similar approach has applied solid-phase microextraction (SPME) and gas chromatography-MS (GC-MS) for investigation of lung cancer volatile biomarkers. Headspace SPME conditions (fiber coating, extraction temperature and extraction time) and desorption conditions were optimized and applied to determination of volatiles in human blood. To find the biomarkers of lung cancer, volatile compounds were investigated in blood of lung cancer patients and controls. Blood concentrations of hexanal and heptanal in lung cancer patients were found to be much higher than those in controls. These preliminary results show that SPME/GC-MS might be a method suitable for investigation of volatile lung cancer markers in human blood *(34)*.

In a third study, breath samples were analyzed by GC and MS to determine alveolar gradients (i.e., the abundance in breath minus the abundance in room air) of volatile organic compounds (C4–C20 alkanes and monomethylated alkanes). Alkanes and monomethylated alkanes are oxidative stress products that are excreted in the breath, the

Table 2
TNM Classification of Lung Cancer

Primary tumor (T)
TX   Primary tumor cannot be assessed, or tumor proven by the presence of malignant cells in sputum or bronchial washings but not visualized by imaging or bronchoscopy
T0   No evidence of primary tumor
Tis   Carcinoma in situ
T1   Tumor 3 cm in greatest dimension, surrounded by lung or visceral pleura, without bronchoscopic evidence of invasion more proximal than the lobar bronchus (i.e., not in the main bronchus)[a]
T2   Tumor with any of the following features of size or extent:
- >3 cm in greatest dimension
- involves main bronchus, 2 cm distal to the carina
- invades the visceral pleura
- associated with atelectasis or obstructive pneumonitis that extends to the hilar region but does not involve the entire lung

T3   Tumor of any size that directly invades any of the following: chest wall (including superior sulcus tumors), diaphragm, mediastinal pleura, parietal pericardium; or tumor in the main bronchus, 2 cm distal to the carina but without involvement of the carina; or associated atelectasis or obstructive pneumonitis of the entire lung
T4   Tumor of any size that invades any of the following: mediastinum, heart, great vessels, trachea, esophagus, vertebral body, carina; or tumor with a malignant pleural or pericar dial effusion,[b] or with satellite tumor nodule(s) within the ipsilateral primary tumor lobe of the lung

Regional lymph nodes (N)
NX   Regional lymph nodes cannot be assessed
N0   No regional lymph node metastasis
N1   Metastasis to ipsilateral peribronchial and/or ipsilateral hilar lymph nodes, and intrapulmonary nodes involved by direct extension of the primary tumor
N2   Metastasis to ipsilateral mediastinal and/or subcarinal lymph node(s)
N3   Metastasis to contralateral mediastinal, contralateral hilar, ipsilateral, or contralateral scalene, or supraclavicular lymph node(s)

Distant metastasis (M)
MX   Presence of distant metastasis cannot be assessed
M0   No distant metastasis
M1   Distant metastasis present[c]

[a] The uncommon superficial tumor of any size with its invasive component limited to the bronchial wall, which may extend proximal to the main bronchus, is also classified T1.

[b] Most pleural effusions associated with lung cancer are due to tumor. However, there are a few patients in whom multiple cytopathologic examinations of pleural fluid show no tumor. In these cases, the fluid is nonbloody and is not an exudate. When these elements and clinical judgment dictate that the effusion is not related to the tumor, the effusion should be excluded as a staging element, and the patient's disease should be staged T1, T2 or T3. Pericardial effusion is classified according to the same rules.

[c] Separate metastatic tumor nodule(s) in the ipsilateral nonprimary tumor lobe(s) of the lung also is classified M1.

catabolism of which may be accelerated by polymorphic cytochrome p450-mixed oxidase enzymes that are induced in patients with lung cancer. Eighty-seven lung cancer patients and 41 healthy volunteers were compared. A predictive model employing nine volatile organic compounds identified primary lung cancer patients with a sensitivity of 90% and a specificity of 83%. The stratification of patients by tobacco smoking status, histologic type of cancer and tumor, node, and metastasis (TNM) stage of cancer (Table 2) revealed no marked effects. Thus, abnormal breath test findings might achieve sufficient sensitivity and specificity to be considered as a screen for lung cancer in a high-risk population such as adult smokers *(35)*.

3. FUNCTIONAL PROTEOMICS STUDIES

For being interpreted correctly, expression proteomics studies need to be complemented by functional investigations. In contrast to the genome, the proteome is a dynamic entity: not only the proteome of a lung cancer cell is different from its normal control, but it is changing over time. Thus, better than genomic studies, proteomics studies allow the investigation of environmental factors over time, and can be used for synthetic analysis of signalling pathways, and of complex protein networks (e.g., the redox system). In the human patient, proteomics technologies can be used for responder profiling studies and for toxicological investigations.

3.1. Signaling Pathways

As seen above, expression proteomics studies have repeatedly highlighted modifications of several kinases and cell cycle proteins in lung cancer. Proteomics studies have been applied to evaluate specifically the degree of phosphorylation of selected proteins in lung cancer. For example, three phosphorylated forms and one unphosphorylated form of the oncoprotein 18 (Op18) were identified and found to be overexpressed in a large group of lung adenocarcinomas as compared with uninvolved lung samples, using quantitative 2D-PAGE analysis with confirmation by MS and two-dimensional Western blot analysis expression. The percentage of phosphorylated to total Op18 protein isoforms increased from 3.2% in normal lung to 7.9% in lung tumors. Both the phosphorylated and unphosphorylated Op18 proteins were significantly increased in poorly differentiated tumors as compared with moderately or well-differentiated lung adenocarcinomas ($p < 0.03$), suggesting that up-regulated expression of Op18 reflects a poor differentiation status and higher cell proliferation rates. The increased expression of Op18 protein was significantly correlated with its mRNA level indicating that increased transcription likely underlies elevated expression of Op18. Indeed, the overexpression of Op18 proteins in poorly differentiated lung adenocarcinomas and the elevated expression of the phosphorylated forms of Op18 might offer a new target for drug-directed therapy *(36)*.

The *C-CRK* gene, cellular homolog of the avian v-crk oncogene, encodes two alternatively spliced adaptor signaling proteins, CRKI (28 kDa) and CRKII (40 kDa). Both CRKI and CRKII have been shown to activate kinase signaling and anchorage-independent growth in vitro, and CRKI transformed cells readily form tumors in nude mice. In lung adenocarcinoma, C-CRK mRNA expression was increased in more advanced, larger and poorly differentiated tumors and in tumors from patients demonstrating poor survival. A significant increase in levels of the CRKI oncoprotein and the phosphorylated isoform of CRKII was observed in tumors. No difference in protein level was evident between stages. Concordant with mRNA expression, CRKI and CRKII were increased in poorly differentiated tumors ($p < 0.05$). CRK immunohistochemical analysis of tumor tissue arrays using the same tumor series also demonstrated increased abundance of nuclear and cytoplasmic CRK in more proliferative tumors. This study provided the first quantitative analysis of discrete CRKI and CRKII protein isoforms in human lung tumors and provided evidence that the C-CRK proto-oncogene may foment a more aggressive phenotype in lung cancers *(37)*.

3.2. Redox System

Alterations of the biology of nitric oxide (NO) metabolites and nitration of proteins may contribute to the mutagenic processes and promote carcinogenesis. Cellular pro-oxidant state promotes cells to neoplastic growth, in part, because of modification of proteins and their functions. Reactive nitrogen species formed from NO or its metabolites can lead to protein tyrosine nitration, which is elevated in over 50 diseases, including lung cancer. Protein nitration may be unique among post-translational modifications in its dependency on reactivity of tyrosine residues in the protein target instead of specific sequence motifs or protein–protein interactions. Tyrosine nitration can cause either gain or loss of protein function *(38)*. In lung cancer, proteomics studies (2D-PAGE, MS) allowed to identify more than 25 proteins modified (degree of nitration and alteration of the protein nitration). These modified proteins included metabolic enzymes, structural proteins and proteins involved in prevention of oxidative damage *(39)*.

3.3. Toxicology Studies

Particulate pollutants, such as diesel exhaust particles (DEPs), may lead to a worsening of various lung diseases, in particular the asthmatic condition. When RAW 264.7 cell lines were treated with DEP chemicals, the induction of oxidative stress was accompanied by 53 newly expressed proteins including antioxidant enzymes, pro-inflammatory components and products of intermediary metabolism. Utilizing 2D-PAGE, anti-nitrotyrosine immunoblotting, and MS led to the identification of an additional 10 nitrotyrosine modified proteins, including oxidative stress proteins involved in intermediary metabolism (e.g., GAPDH and enolase), antioxidant defense (e.g., MnSOD) and inhibition of proteosomal activity (e.g., Hsp 90alpha) *(40)*.

3.4. Responder Profiling

Responder profiling may be useful in tailoring chemotherapeutic protocols to individual tumors. Cisplatin-based chemotherapy (CDDP) is the established adjuvant treatment after surgery for NSCLC and was reported to yield 5–15% improvement in 5-year survival compared to surgery alone, but can only be achieved with substantial toxic effects. Thus, preselection of good responders would be very helpful for allowing individualized therapeutic strategies. Proteomics techniques (2D-PAGE) have been applied for identifying responders to CDDP therapy. Protein expression intensity was compared between parent strains (H69 and PC14 lung cancer cultured cells) and resistant strains against CDDP, and differentially expressed polypeptides were identified as reticulocalbin (RCN) and glutathione-S-transferase-pi (GST-pi). When the relationship between protein expression associated with CDDP resistance and the clinical effects of platinum-based postoperative adjuvant chemotherapy was tested on 126 surgically resected NCLC materials, RCN-positive cases showed a statistically significant better disease-free survival only in the cases receiving postoperative CDDP chemotherapy after curative resection ($p = 0.007$) *(41)*. Interestingly, response to CDDP chemotherapy was known for 10 years to be significantly related to expression of GST-pi in NSCLC patients *(42–44)* so that the present example can serve as a proof of concept for applying proteomics technologies for responder profiling.

3.5. Discovery of Novel Cellular Processes and Networks

In lung adenocarcinoma, using oligonucleotide arrays, correlation between mRNA expression levels of vascular endothelial growth factor (VEGF) and all other genes ($n = 4966$) was used to identify other biologic processes that may be associated with increased VEGF expression. VEGF and IGFBP3 mRNA were found to be overexpressed in bronchial-derived lung adenocarcinomas, and expression was decreased in well-differentiated lung adenocarcinomas. 2D-PAGE was used to analyze the protein expression profiles of VEGF and IGFBP3 isoforms and revealed several protein isoforms. Forty genes were identified as the most significantly associated with VEGF expression, 17 of which were also associated with IGFBP3 and 12 were known to be induced through the HIF1 pathway. Among other highly correlated genes, several, including bradykinin receptor B2, suggest additional cellular processes that were not previously known to be associated with VEGF expression in lung adenocarcinoma *(45)*.

3.6. Identification of Tumor-Associated Antigens

The identification of circulating tumor antigens or their related autoantibodies provides a means for early cancer diagnosis as well as leads for therapy. Research over the past decade has resulted in some reports on the presence of autoantibodies against disease-related proteins such as annexins I and II *(46)*, recoverin *(47)* and protein gene product 9.5 *(48)* in the sera of patients with lung cancer. Comparsion of 2D-PAGE/Western blot/electrochemi-luminescence (ECL) detection revealed distinct distributions of antibodies in the sera of lung adenocarcinoma, tuberculosis and healthy subjects and allowed the detection of 16 protein spots in cancer patients, including alpha-enolase, chaperonin and others. An antibody against alpha-enolase was observed in three of five patients with adenocarcinoma, but not in patients with tuberculosis and not in healthy subjects, suggesting some specificity of alpha-enolase as a marker of lung adenocarcinoma *(49)*.

Peroxiredoxin-I (Prx-I) autoantibody and circulating antigen might be potential biomarkers for use in serological diagnosis of NSCLC. In eukaryotic cells, peroxiredoxins are both antioxidants and regulators of H_2O_2-mediated signaling. Prx-I is overexpressed in NSCLC tissue, and, using Western blotting, it was found that 47% of NSCLC patients tested had circulating autoantibodies against Prx-I, whereas such activity was detected in 8% of healthy subjects. Prx-I itself was detected in the sera from 34% of NSCLC patients but in only 2% of controls. Moreover,

17% of NSCLC sera were positive to both Prx-I antibody and antigen but none in control sera *(50)*.

3.7. Correlation with Clinical Outcome

Morphologic assessment of lung tumors is informative but insufficient to adequately predict patient outcome. Prognosis is currently defined by determining the extension of disease at time of diagnosis, and the internationally accepted TNM classification (Table 2) serves as a basis for prognostic assessment and therapy recommendations. However, this classification does not include data from molecular profiling, which might be useful to better predict prognosis. In lung cancer, protein profiling was shown to predict survival in UICC stage I tumor patients: out of total of 682 individual protein spots detected in 90 lung adenocarcinoma, 90 proteins were quantified, and 20 were selected by Cox modeling as survival-associated proteins. Thirty-three survival-associated proteins could be identified by MS, and expression of 12 candidate proteins was confirmed as tumor derived with immunohistochemical analysis and tissue microarrays. Oligonucleotide microarray results from both the same tumors and from an independent study showed mRNAs associated with survival for 11 of 27 encoded genes. Combined analysis of protein and mRNA data revealed 11 components of the glycolysis pathway as associated with poor survival. Among these candidates, phosphoglycerate kinase 1 was associated with survival in the protein study, in both mRNA studies and an independent validation set of 117 adenocarcinomas and squamous lung tumors using tissue microarrays. Elevated levels of phosphoglycerate kinase 1 in the serum were also significantly correlated with poor outcome in a validation set of 107 patients with lung adenocarcinomas using ELISA analysis. These proteomics studies identified new prognostic biomarkers in lung adenocarcinoma and indicated that protein expression profiles can predict the outcome of early stage tumors *(51)*.

MS peptide fingerprints obtained directly from small amounts of fresh frozen lung tumor tissue (51) could be used to accurately distinguish between patients who had poor prognosis (median survival 6 months, $n = 25$) and those who had good prognosis (median survival 33 months, $n = 41$, $p < 0.0001$) after surgical resection in curative intent *(52)*. Taken together, these studies suggest that proteomic patterns obtained with 2D-PAGE or directly from fresh frozen lung tumor tissue could be used to accurately predict survival in resected NSCLC.

4. CONCLUSION

Proteomics technologies are increasingly applied in translational lung cancer research. Although they are not in use for long, promising results have already been achieved in the diagnostic, prognostic as well as in the therapeutic areas. Proteomics is a rapidly evolving field, and novel developments, in particular technological improvements such as gel-free MS, but also the recent publication of the first results of the collaborative effort of the Human Plasma Proteome Project, might increase significantly the pace of research in the coming years. However, careful sample preparation techniques, analysis of sufficient numbers of samples and a close link between bench and bedside will remain limiting factors in translating technological progress into useful diagnostic and therapeutic tools.

REFERENCES

1. Granville CA, Dennis PA. An overview of lung cancer genomics and proteomics. Am J Respir Cell Mol Biol 2005 Mar;32(3):169–76. Review.
2. Li C, Zhan X, Li M, Wu X, Li F, Li J, Xiao Z, Chen Z, Feng X, Chen P, Xie J, Liang S. Proteomic comparison of two-dimensional gel electrophoresis profiles from human lung squamous carcinoma and normal bronchial epithelial tissues. Genomics Proteomics Bioinformatics 2003 Feb;1(1):58–67.
3. Li C, Chen Z, Xiao Z, Wu X, Zhan X, Zhang X, Li M, Li J, Feng X, Liang S, Chen P, Xie JY. Comparative proteomics analysis of human lung squamous carcinoma. Biochem Biophys Res Commun 2003 Sep 12;309(1):253–60.
4. Wu X, Xiao Z, Chen Z, Li C, Li J, Feng X, Yi H, Liang S, Chen P. Differential analysis of two-dimension gel electrophoresis profiles from the normal-metaplasia-dysplasia-carcinoma tissue of human bronchial epithelium. Pathol Int 2004 Oct;54(10):765–73.
5. Zhukov TA, Johanson RA, Cantor AB, Clark RA, Tockman MS. Discovery of distinct protein profiles specific for lung tumors and pre-malignant lung lesions by SELDI mass spectrometry. Lung Cancer 2003 Jun;40(3):267–79.
6. Sueoka E, Sueoka N, Goto Y, Matsuyama S, Nishimura H, Sato M, Fujimura S, Chiba H, Fujiki H. Heterogeneous nuclear ribonucleoprotein B1 as early cancer biomarker for occult cancer of human lungs and bronchial dysplasia. Cancer Res 2001 Mar 1;61(5):1896–902.
7. Tominaga M, Sueoka N, Irie K, Iwanaga K, Tokunaga O, Hayashi S, Nakachi K, Sueoka E. Detection and discrimination of preneoplastic and early stages of lung adenocarcinoma using hnRNP B1 combined with the cell cycle-related markers p16, cyclin D1, and Ki-67. Lung Cancer 2003 Apr;40(1):45–53.
8. Matsuyama S, Goto Y, Sueoka N, Ohkura Y, Tanaka Y, Nakachi K, Sueoka E. Heterogeneous nuclear ribonucleoprotein B1 expressed in esophageal squamous cell carcinomas as a new biomarker for diagnosis. Jpn J Cancer Res 2000 Jun;91(6):658–63.
9. Chen G, Gharib TG, Huang CC, Thomas DG, Shedden KA, Taylor JM, Kardia SL, Misek DE, Giordano TJ, Iannettoni MD, Orringer MB, Hanash SM, Beer DG.

Proteomic analysis of lung adenocarcinoma: identification of a highly expressed set of proteins in tumors. Clin Cancer Res 2002 Jul;8(7):2298–305.
10. Gharib TG, Chen G, Wang H, Huang CC, Prescott MS, Shedden K, Misek DE, Thomas DG, Giordano TJ, Taylor JM, Kardia S, Yee J, Orringer MB, Hanash S, Beer DG. Proteomic analysis of cytokeratin isoforms uncovers association with survival n lung adenocarcinoma. Neoplasia 2002 Sep–Oct;4(5):440–8.
11. Seike M, Kondo T, Fujii K, Okano T, Yamada T, Matsuno Y, Gemma A, Kudoh S, Hirohashi S. Proteomic signatures for histological types of lung cancer. Proteomics 2005 Jul;5(11):2939–48.
12. Craven RA, Banks RE. Laser capture microdissection and proteomics: possibilities and limitation. Proteomics 2001 Oct;1(10):1200–4. Review.
13. Reymond MA, Sanchez JC, Hughes GJ, Gunther K, Riese J, Tortola S, Peinado MA, Kirchner T, Hohenberger W, Hochstrasser DF, Kockerling F. Standardized characterization of gene expression in human colorectal epithelium by two-dimensional electrophoresis. Electrophoresis 1997 Dec;18(15):2842–8.
14. Zhukov TA, Johanson RA, Cantor AB, Clark RA, Tockman MS. Discovery of distinct protein profiles specific for lung tumors and pre-malignant lung lesions by SELDI mass spectrometry. Lung Cancer 2003 Jun;40(3):267–79.
15. Yanagisawa K, Shyr Y, Xu BJ, Massion PP, Larsen PH, White BC, Roberts JR, Edgerton M, Gonzalez A, Nadaf S, Moore JH, Caprioli RM, Carbone DP. Proteomic patterns of tumour subsets in non-small-cell lung cancer. Lancet 2003 Aug 9;362(9382):433–9.
16. Luche S, Santoni V, Rabilloud T. Evaluation of nonionic and zwitterionic detergents as membrane protein solubilizers in two-dimensional electrophoresis. Proteomics 2003 Mar;3(3):249–53.
17. Wu XY, Xiao ZQ, Chen ZC, Li C, Li JL, Feng XP, Yi H, Li MY. [Improvement of protein preparation methods for bronchial epithelial tissues and establishment of 2-DE profiles from carcinogenic process of human bronchial epithelial tissues] Zhong Nan Da Xue Xue Bao Yi Xue Ban 2004 Aug;29(4):376–81. Chinese.
18. Maciel CM, Paschoal ME, Kawamura MT, Carvalho Mda G. Serum protein profiling of lung cancer patients. J Exp Ther Oncol 2004 Dec;4(4):327–34.
19. Yang SY, Xiao XY, Zhang WG, Zhang LJ, Zhang W, Zhou B, Chen G, He DC. Application of serum SELDI proteomic patterns in diagnosis of lung cancer. BMC Cancer 2005 Jul 20;5:83.
20. Xiao XY, Tang Y, Wei XP, He DC. A preliminary analysis of non-small cell lung cancer biomarkers in serum. Biomed Environ Sci 2003 Jun;16(2):140–8.
21. Ebert MP, Meuer J, Wiemer JC, Schulz HU, Reymond MA, Traugott U, Malfertheiner P, Rocken C. Identification of gastric cancer patients by serum protein profiling. J Proteome Res 2004 Nov–Dec;3(6):1261–6.
22. Fujii K, Nakano T, Kanazawa M, Akimoto S, Hirano T, Kato H, Nishimura T. Clinical-scale high-throughput human plasma proteome analysis: lung adenocarcinoma. Proteomics 2005 Mar;5(4):1150–9.
23. Tyan YC, Wu HY, Lai WW, Su WC, Liao PC. Proteomic profiling of human pleural effusion using two-dimensional nano liquid chromatography tandem mass spectrometry. J Proteome Res 2005 Jul–Aug;4(4):1274–86.
24. Tyan YC, Wu HY, Su WC, Chen PW, Liao PC. Proteomic analysis of human pleural effusion. Proteomics 2005 Mar;5(4):1062–74.
25. Wattiez R, Falmagne P. Proteomics of bronchoalveolar lavage fluid. J Chromatogr B Analyt Technol Biomed Life Sci 2005 Feb 5;815(1–2):169–78.

26. Noel-Georis I, Bernard A, Falmagne P, Wattiez R. Proteomics as the tool to search for lung disease markers in bronchoalveolar lavage. Dis Markers 2001;17(4):271–84. Review.
27. Bell DY, Hook GE. Pulmonary alveolar proteinosis: analysis of airway and alveolar proteins. Am Rev Respir Dis 1979 Jun;119(6):979–90.
28. Lindahl M, Stahlbom B, Svartz J, Tagesson C. Protein patterns of human nasal and bronchoalveolar lavage fluids analyzed with two-dimensional gel electrophoresis. Electrophoresis 1998 Dec;19(18):3222–9.
29. Wattiez R, Hermans C, Bernard A, Lesur O, Falmagne P. Human bronchoalveolar lavage fluid: two-dimensional gel electrophoresis, amino acid microsequencing and identification of major proteins. Electrophoresis 1999 Jun;20(7):1634–45.
30. Wattiez R, Hermans C, Cruyt C, Bernard A, Falmagne P. Human bronchoalveolar lavage fluid protein two-dimensional database: study of interstitial lung diseases. Electrophoresis 2000 Jul;21(13):2703–12.
31. Noel-Georis I, Bernard A, Falmagne P, Wattiez R. Database of bronchoalveolar lavage fluid proteins. J Chromatogr B Analyt Technol Biomed Life Sci 2002 May 5;771(1–2):221–36. Review.
32. Scheideler L, Manke HG, Schwulera U, Inacker O, Hammerle H. Detection of nonvolatile macromolecules in breath. A possible diagnostic tool? Am Rev Respir Dis 1993 Sep;148(3):778–84.
33. Machado RF, Laskowski D, Deffenderfer O, Burch T, Zheng S, Mazzone PJ, Mekhail T, Jennings C, Stoller JK, Pyle J, Duncan J, Dweik RA, Erzurum SC. Detection of lung cancer by sensor array analyses of exhaled breath. Am J Respir Crit Care Med 2005 Jun 1;171(11):1286–91.
34. Deng C, Zhang X, Li N. Investigation of volatile biomarkers in lung cancer blood using solid-phase microextraction and capillary gas chromatography-mass spectrometry. J Chromatogr B Analyt Technol Biomed Life Sci 2004 Sep 5;808(2):269–77.
35. Phillips M, Cataneo RN, Cummin AR, Gagliardi AJ, Gleeson K, Greenberg J, Maxfield RA, Rom WN. Detection of lung cancer with volatile markers in the breath. Chest 2003 Jun;123(6):2115–23.
36. Chen G, Wang H, Gharib TG, Huang CC, Thomas DG, Shedden KA, Kuick R, Taylor JM, Kardia SL, Misek DE, Giordano TJ, Iannettoni MD, Orringer MB, Hanash SM, Beer DG. Overexpression of oncoprotein 18 correlates with poor differentiation in lung adenocarcinomas. Mol Cell Proteomics 2003 Feb;2(2):107–16.
37. Miller CT, Chen G, Gharib TG, Wang H, Thomas DG, Misek DE, Giordano TJ, Yee J, Orringer MB, Hanash SM, Beer DG. Increased C-CRK proto-oncogene expression is associated with an aggressive phenotype in lung adenocarcinomas. Oncogene 2003 Sep 11;22(39):7950–7.
38. Aulak KS, Miyagi M, Yan L, West KA, Massillon D, Crabb JW, Stuehr DJ. Proteomic method identifies proteins nitrated in vivo during inflammatory challenge. Proc Natl Acad Sci USA 2001 Oct 9;98(21):12056–61.
39. Masri FA, Comhair SA, Koeck T, Xu W, Janocha A, Ghosh S, Dweik RA, Golish J, Kinter M, Stuehr DJ, Erzurum SC, Aulak KS. Abnormalities in nitric oxide and its derivatives in lung cancer. Am J Respir Crit Care Med 2005 Sep 1;172(5):597–605.
40. Xiao GG, Nel AE, Loo JA. Nitrotyrosine-modified proteins and oxidative stress induced by diesel exhaust particles. Electrophoresis 2005 Jan;26(1):280–92.

41. Hirano T, Kato H, Maeda M, Gong Y, Shou Y, Nakamura M, Maeda J, Yashima K, Kato Y, Akimoto S, Ohira T, Tsuboi M, Ikeda N. Identification of postoperative adjuvant chemotherapy responders in non-small cell lung cancer by novel biomarker. Int J Cancer 2005 Nov 10;117(3):460–8.
42. Bai F, Nakanishi Y, Kawasaki M, Takayama K, Yatsunami J, Pei XH, Tsuruta N, Wakamatsu K, Hara N. Immunohistochemical expression of glutathione S-transferase-Pi can predict chemotherapy response in patients with nonsmall cell lung carcinoma. Cancer 1996 Aug 1;78(3):416–21.
43. Hida T, Kuwabara M, Ariyoshi Y, Takahashi T, Sugiura T, Hosoda K, Niitsu Y, Ueda R. Serum glutathione S-transferase-pi level as a tumor marker for non-small cell lung cancer. Potential predictive value in chemotherapeutic response. Cancer 1994 Mar 1;73(5):1377–82.
44. Inoue T, Ishida T, Sugio K, Maehara Y, Sugimachi K. Glutathione S transferase Pi is a powerful indicator in chemotherapy of human lung squamous-cell carcinoma. Respiration 1995;62(4):223–7.
45. Gharib TG, Chen G, Huang CC, Misek DE, Iannettoni MD, Hanash SM, Orringer MB, Beer DG. Genomic and proteomic analyses of vascular endothelial growth factor and insulin-like growth factor-binding protein 3 in lung adenocarcinomas. Clin Lung Cancer 2004 Mar;5(5):307–12.
46. Brichory FM, Misek DE, Yim AM, Krause MC, Giordano TJ, Beer DG, Hanash SM. An immune response manifested by the common occurrence of annexins I and II autoantibodies and high circulating levels of IL-6 in lung cancer. Proc Natl Acad Sci USA 2001 Aug 14;98(17):9824–9.
47. Jankowska R, Witkowska D, Porebska I, Kuropatwa M, Kurowska E, Gorczyca WA. Serum antibodies to retinal antigens in lung cancer and sarcoidosis. Pathobiology 2004;71(6):323–8.
48. Brichory F, Beer D, Le Naour F, Giordano T, Hanash S. Proteomics-based identification of protein gene product 9.5 as a tumor antigen that induces a humoral immune response in lung cancer. Cancer Res 2001 Nov 1;61(21):7908–12.
49. Ueda K. [Proteome analysis of autoantibodies in sera of patients with cancer] Rinsho Byori 2005 May;53(5):437–45.
50. Chang JW, Lee SH, Jeong JY, Chae HZ, Kim YC, Park ZY, Yoo YJ. Peroxiredoxin-I is an autoimmunogenic tumor antigen in non-small cell lung cancer. FEBS Lett 2005 May 23;579(13):2873–7.
51. Chen G, Gharib TG, Wang H, Huang CC, Kuick R, Thomas DG, Shedden KA, Misek DE, Taylor JM, Giordano TJ, Kardia SL, Iannettoni MD, Yee J, Hogg PJ, Orringer MB, Hanash SM, Beer DG. Protein profiles associated with survival in lung adenocarcinoma. Proc Natl Acad Sci USA 2003 Nov 11;100(23):13537–42.
52. Yanagisawa K, Shyr Y, Xu BJ, Massion PP, Larsen PH, White BC, Roberts JR, Edgerton M, Gonzalez A, Nadaf S, Moore JH, Caprioli RM, Carbone DP. Proteomic patterns of tumour subsets in non-small-cell lung cancer. Lancet 2003 Aug 9;362(9382):433–9.
53. World Health Organization. Histological Typing of Lung Tumors. 2nd ed. Geneva, Switzerland: WHO; 1981.

8 Proteomic Strategies of Therapeutic Individualization and Target Discovery in Acute Myeloid Leukemia

Bjørn Tore Gjertsen and Gry Sjøholt

CONTENTS

1. INTRODUCTION
2. PROTEOMICS
3. ANALYSIS OF CLINICAL SAMPLES
4. AML DIAGNOSTICS
5. PHARMACOPROTEOMICS IN AML
6. CONCLUDING REMARKS

SUMMARY

Acute myeloid leukemia (AML) is an aggressive hematological malignancy characterized by accumulating myeloid precursor cells in the bone marrow, with approximately 2–3 months of 50% survival if left untreated. If current treatment modalities are employed, the 5-year overall survival hardly exceeds 50%. Cytogenetics and molecular diagnostics guide the clinician to individualized therapy in a minority of AML, achieving long-term survival above 70% of these cases. However, approximately half of the AML patients have no risk stratifying features, and early reports indicate that proteomic approaches may be utilized for disease classification as well as

From: *Cancer Drug Discovery and Development
Cancer Proteomics: From Bench to Bedside*
Edited by: S. S. Daoud © Humana Press Inc., Totowa, NJ

development of novel biomarkers related to prognosis, diagnosis and choice of therapeutic regimes. Proteomics, here defined as the analysis of all proteins in a cell, in a cell compartment or in a signaling pathway, has probably its greatest potential in directly investigating pathways that are targeted by small molecules or therapeutic antibodies. The major methodological challenges include detection sensitivity in a limited clinical material. In this chapter, we will discuss pharmacoproteomic studies of drugs regulating epigenetics, gene transcription, protein degradation and signal transduction. Proteomic strategies may be feasible to elucidate mechanisms of drug resistance and the cause of disease relapses. An important emerging field is the use of proteomics in monitoring biological therapeutical modalities like hematopoietic stem cell transplantation and its graft-versus-leukemia effect and the graft-versus-host reactions. These proteomic applications indicate new avenues in AML diagnostics, individualized therapy design and therapy response monitoring for the clinician

Key Words: Biomarkers; pharmacoproteomics; all-trans retinoic acid; tyrosine kinase inhibitors; histone deacetylase inhibitors; proteasome inhibitors

1. INTRODUCTION

Development of acute leukemia involves block in differentiation of the hematopoietic progenitors. This incomplete maturation of leukemic blasts is usually accompanied with extensive cell proliferation, resulting in cell accumulation in bone marrow, peripheral blood, and eventually at ectopic locations like in the liver, in the spleen and in the central nervous system *(1)*. The acute leukemias are grossly divided into acute lymphoblastic leukemia (ALL) and acute myeloid leukemia (AML) according to the cell lineage of origin. A recent WHO classification. of myeloid malignancies is classifying AML based on the presence of recurrent chromosomal translocations and on morphological features of the leukemia cells if these translocations are missing *(2)*. Importantly—and with few exceptions in contrast to the morphological classification—these recurrent genetic aberrations appear predictive for the prognosis of patients who receive chemotherapy *(3)*. Gene expression analysis appears to reflect the WHO classification and may additionally provide an ability to risk stratify patients without known genetic aberrancies *(4–6)*. Recent reports indicate that a proteomic approach may also reproduce the WHO disease classification *(7)*.

Conventional induction therapy of acute leukemia is chemotherapy and in selected cases hematopoietic stem cell transplantation in first remission *(1,8)*. Overall, 5-year survival in large study materials of AML hardly exceeds 50%, and for patients above 60 years of age, the long-term survival is drastically lower. The majority of AML patients is in these older age groups and is usually excluded for the intensive therapy due to increased risk for fatal toxicities. The potential of individualization of chemotherapy is illustrated by therapeutic use of all-trans retinoic acid (ATRA) in AML with altered retinoic acid receptor (RAR) *(9)*. ATRA combined with conventional chemotherapy has improved long-term survival from approximately 50–70% in patients with fusion proteins of the RARα. The concept of targeted therapy is elegantly demonstrated in chronic myeloid leukemia (CML), a myeloid stem cell disease characterized by fusion protein of the serine/threonine kinase Bcr (Breakpoint cluster region) and the Abelson tyrosine kinase (Bcr-Abl chimeric protein). The selective kinase inhibitor imatinib mesylate has been successful in disease control of CML, but does probably not cure the majority of patients due to development of Abelson kinase (Abl) point mutations *(10)*. Motivated by the need of novel therapeutic approaches in AML, we will in the following sections discuss the potential use of proteomics in development of individualized therapy in AML, illustrated by protein studies of hematological malignancies.

2. PROTEOMICS

All the proteins involved in a particular signaling pathway, organelle, cell, tissue, organ or organism are considered as proteomes *(11–13)*, and proteomics can be defined as comprehensive identification, quantification and characterization of the multiple proteins in proteomes. Gene expression studies with DNA microarrays do not necessarily reflect the number of proteins in a cell, or different posttranslational modifications present in a protein *(14)*. Consequently, determination of protein expression levels and protein modifications appear to be necessary and complimentary techniques to mRNA studies to obtain a more complete picture of a cancer cell *(15)*. Technological improvements during the last decade allow the biomedical researcher to study the numerous proteins in a proteome. Because of the huge amount of data available in a proteome, selection of important proteins may be necessary to obtain useful information.

2.1. Separation of Proteins and Peptides

Proteomics methodology depends on separation of complex protein mixtures from lysed cells *(16)*. A frequently used method is two-dimensional polyacrylamide gel electrophoresis (2DE), separating the proteins based on molecular charge and molecular weight *(17,18)*. An advantage of comparative 2DE analysis is its capability to visualize differences in protein patterns due to charge shifts (e.g., due to phosphorylation or acetylation) and changes in molecular weight (e.g., due to cleavage or adduction) of distinct protein spots. Critically, dynamic range, detection limits, protein weight discrimination and protein solubility are issues that limit the utilization of 2DE. Capillary separation methodologies eliminate some of the problems encountered in 2DE, and multidimensional liquid chromatography (LC) is increasingly employed as protein and peptide separation techniques in proteomics *(19–21)*. The high-pressure liquid chromatography (HPLC) system used most commonly for peptide separation inline with electrospray ionization mass spectrometry (MS) is reverse-phase liquid chromatography. Both proteins and peptides can also be separated by size exclusion chromatography, capillary electrophoresis, isoelectric focusing and ion exchange chromatography *(19–21)*.

2.2. Mass Spectrometric Technology

Proteins separated by 2DE gels or by LC fractionation are identified by MS. MS rapidly and accurately measures the molecular weight of peptides and proteins *(16)*. The ionization source of a mass spectrometer converts and transfers molecules (analytes) into gas-phase ions *(22,23)*. On the basis of variations in the mass/charge ratio (m/z), these ions migrate with differences in time-of-flight as measured by a detector. In the resulting MS spectra, each peak represents one molecule with a specific m/z. Proteins can be digested with a protease with specific sequence demands, and consequently, the size of the resulting peptides and the corresponding peptide mass fingerprint (PMF) spectra can be predicted. As a result of the Human Genome Project and easily accessible protein sequence databases, researchers can compare obtained PMF spectra with predicted spectra of proteins in the databases. Consequently, when PMFs are matched, the protein can be identified. When performing MS/MS, the peptide ion of a certain peak is further fragmented, and by the resulting fragment spectrum, the amino acid sequence of a peptide can be predicted.

MS can also be used to quantify peptide abundances through comparing peak intensities in peptide profiles, a field in proteomics that has evolved considerably during the last years *(24)*. Proteins or peptides from different samples are modified with specific isotopes or linker molecules with different mass. Prior to LC-MS/MS analyses, the labeled samples (for instance treated and untreated) are pooled. Examples of different MS-based methods to compare relative ratios of peptide quantities are isotope-coded affinity tag *(25)*, isobaric tags for relative and absolute quantification *(26)*, ^{18}O-incorporation *(27)* and stable isotope labeling with amino acids in cell culture (SILAC) *(28)*. Peptides represented in two samples are differently labeled before analysis under identical conditions for HPLC and MS. Due to labeling with different isotopes, identical peptides are represented as double peaks in the MS spectra representing small shifts in m/z ratio. Comparing signal strengths of shifted MS peaks in all of these methods allow relative quantification of peptides to be identified by MS/MS. Another MS-based method for absolute quantification (AQUA) has also been established *(29,30)*. The AQUA method relies on synthetic internal standard peptides that are introduced at a known concentration to cell lysates during digestion.

An alternative commercialized option for protein profiling analysis is the surface-enhanced laser desorption/ionization (SELDI) approach *(16)*. SELDI analyses are run with an adsorptive surface chemistry, for example, cation/anion exchange material or hydrophobic surface; thereafter, relatively small proteins captured by matrix solution are profiled by MS-based technology *(31–35)*. SELDI is most effective on small proteins (<30 kDa) and typically function as a biomarker method as the protein identification can be difficult and time consuming.

2.3. Multiplexed Protein Assays in Cell Extracts or Intact Cells

Identification of a specific protein through proteomic screens may be exploited in development of more time and cost saving methodology, in particular for further testing in numerous patient materials *(36)*. Techniques based on antibody recognition of known proteins in cell extracts, for instance in immunoblots, are frequently utilized as quantitative protein assays. Multiplexed immunoassays can be used both as screening methods and to verify results from other screening methods in quantitative proteomics. A constantly increasing number of antibodies are commercially available, and to systemize the increasing amount of antibody-related information, a HUPO-supported database, Human Protein Atlas (http://www.proteinatlas.org/), has been established to give a complete overview of all available antibodies binding to human proteins. Multiplex analysis of proteins can be performed in intact cells or in cell extracts. Because multiplexed protein assays with standardized procedures are flexible, rapid and require small sample volumes, these techniques might be an increasing part of leukemia diagnostics in the future.

Intact cells can be analyzed in multiplex antibody based flow cytometry assays, successfully performed on AML patients material where the phosphorylated form of several proteins (Stat-1, Stat-3, Stat-5, Stat-6, p38 and Erk1/2) was studied before and 15 min after growth factor stimulation *(15)*. These signaling nodes are part of signal transduction pathways that modulate gene expression of proteins involved in cell death regulation and proliferation, and the activation state of these nodes may therefore reflect mechanisms involved in chemoresistance. Automated machine learning was applied to make signaling profiles achieving to determine signaling maps from individual patient samples *(37)*. Multiplexed flow cytometry may allow efficient analysis of a limited sample material. However, a limitation is the availability of verified antibodies that are compatible with flow cytometry, but these numbers are expected to rapidly increase *(37)*. An important advantage of flow cytometry-based protein analysis on intact cells may include information about various cancer cell clones in the sample. This may be particularly important in hematological malignancies where it is suggested that a subset of leukemia stem cells are chemoresistant and unavailable for effective eradication by conventional therapy *(38)*.

Enzyme-linked immunoabsorbant assay (ELISA) has been the standard method to measure protein quantities in solutions. In its most common form, a plate-bound antibody captures a specific protein in the sample solution. A second antibody recognizes another epitope on the protein, forming a sandwich. The secondary antibody can be directly linked to an enzyme or a fluorescence dye, or it can be recognized by a third antibody that is linked to an enzyme or a fluorescence dye. The colorimetric or chemiluminescence product of the enzyme or fluorescence intensity from dye can be measured to quantify the amount of captured proteins. Traditionally, ELISAs are performed on 96-well plates, testing the same antibody on one plate. However, with novel methodology, several antibodies can be spotted in the same well, and multiple sample proteins can be quantified on the same plate *(39)*.

Simultaneous analysis of multiple proteins in cellular extracts or biological fluids can be performed in a system using microbeads coated with specific antibodies and run through a flow cytometer system *(40)*. The usefulness of this rapid and sensitive method has been demonstrated in a study of phosphoproteomic profiles in lymphoid cell lines *(41)*, using multiplex suspension arrays to investigate phosphorylation dynamics and kinetics in signal transduction pathways. The specificity of the method was well demonstrated by analysis of two phosphorylated kinases, Zap-70 and Syk. High phosphorylation levels of Zap-70 was detected in Jurkat but not in Burkitt lymphoma Ramos B cells, which correlates with high protein expression of Zap-70 in B-cell Ramos Chronic Lymphocytic Leukemia (CLL) patients and a more aggressive disease *(42)* while phosphorylation levels of Syk were elevated in Ramos B cells. These findings were confirmed by complementary techniques like immunoaffinity purification and Western blots.

Protein arrays are a developing technology for quantification and functional analysis in proteomic research *(32,43–47)*. Depending on the field of application, protein arrays can be classified into two categories: (1) arrays for expression proteomics and (2) arrays for functional studies *(32,44)*. In functional studies, the analysis of protein–protein, enzyme–substrate, protein–DNA, protein–oligosaccharide and protein–drug interactions has been described. For expression studies, the protein abundance in a sample has been quantitatively determined by different high-affinity protein-binding techniques; most frequently antibodies are arrayed at high spatial density on a solid support. In contrast to high-density DNA arrays, which facilitate a global view of the transcriptome, current antibody arrays permit the expression analysis of only a limited number of proteins. However, providing only a small number of parameters need to be analyzed; today's protein arrays are an excellent solution for protein profiling approaches. Interestingly, Belov et al. *(48)* have made microarrays of novel monoclonal antibodies for binding to T, B and myeloid leukemia cells. Using an array containing 82 antibodies directed against various leukocyte surface molecules, the immunophenotype was achieved in normal peripheral blood cells and leukemia. The method has been improved by making an array containing 498 monoclonal antibodies to further differentiate the leukocytes *(49)*. This method may be used for clustering analysis, and compared with conventional flow cytometry or immunohistochemistry, more data are collected with the use of less patient material.

3. ANALYSIS OF CLINICAL SAMPLES

A growing number of studies are applying proteomics on clinical samples from leukemia patients *(7,50–54)*. Obviously, sample preparation is critical to the success of proteomic analysis of patient material *(55)*, and the study design must take into consideration experimental variation originating from sample collection procedures and normal biological variability. In general, low abundance proteins with regulatory functions may fall below the detection limit of current proteomic methodology; this is complicating the use of proteomics in clinical studies where the sample material may be limited *(56)*.

Enrichment protocols may increase the relative representation of low abundance proteins in the samples to be analyzed. Two approaches of protein enrichment are subcellular fractionation and protein prefractionation. Prefractionation of posttranslationally modified proteins such as phosphorylated, acetylated and ubiquitinated proteins can be performed by affinity chromatography *(57–59)* and/or immunoprecipitation *(58,60,61)*. Subcellular fractionations are advantageous as specific low abundance proteins can be predominantly present in a cell compartment. Proteomics comparison of various subcellular compartments make it possible to assign cellular localization of a particular protein of interest and thus monitor intracellular protein trafficking *(62)*. The potential of studying differentially expressed proteins in specified cellular compartments is also a major advantage of proteomics in contrast to gene expression studies.

In human plasma, the most abundant protein, albumin, constitute approximately 50% of the total protein mass, and immunoglobulins represent 20–25% of the total protein mass. One strategy to extend the dynamic range of proteome detection is through depletion of major proteins in the sample to be studied. The introduction of affinity chromatography now allows for the selective removal of albumin and immunoglobulins, extending the dynamic range and increasing the coverage and diversity of proteins to be identified in plasma proteomics studies *(63)*. Even if commercially available columns may be imperfect regarding efficiency and specificity, increased protein diversity has been demonstrated by applying chromatography depletion techniques on plasma samples *(64)*. Therefore, we suggest that depletion strategy should be considered when measuring biomarkers in plasma samples of leukemia.

Most clinical samples will contain a mixture of normal cells and cancer cells, and within the cancer cell, heterogeneity will be present. In analysis on total cell extract, this heterogeneity of cell populations may exert significant effects on the results. In AML, the putative leukemia stem cell is present only in low numbers *(65–67)*, whereas current diagnostics are based on karyotypic analysis of relatively unselected leukocytes from the bone marrow or peripheral blood *(2)*. It is currently not known whether risk stratification may be improved if leukemia stem cells were used as the diagnostic basis. An apparent cell clone independence in the karyotypic analysis is supported by gene expression analysis of AML cells collected from peripheral blood or bone marrow, demonstrating similar capacities in determination of risk stratification *(5)*. These results are in contrast to the observable differences in surface molecule expression in bone marrow and peripheral blood *(68,69)*. Similarly, cell population heterogeneity is indicated in signaling node responses of AML *(15)*, which indicates that more detailed prognostics may be obtained by analysis of leukemia cell subsets. Still, AML appear homogenous enough to give clinical useful information from standard cell sampling and total cell extracts. With the availability of patients and relatively easily accessible cancer cell sampling, leukemia may be considered a good model for applying novel methodologies in the study of biological mechanisms of cancer.

4. AML DIAGNOSTICS

Karyotyping, cytochemistry and immunophenotypic analyses are diagnostic methods that distinguish between the different leukemias and subclasses. The recent WHO classification set the criteria for AML to cytological detection of more than 20% myeloid blasts in bone marrow, or presence of recurrent karyotypic aberrancies in myeloid cells of bone marrow or peripheral blood *(2)*.

As early as in 1988, acute leukemia that could not be distinguished by light microscopy and morphological criteria was clustered by immunophenotyping and DNA analysis *(70)*. Immunophenotypically, approximately 60% of acute leukemia samples expressed one or two myeloid antigens, whereas remaining cases had a complex phenotype with expression of both myeloid and lymphoid antigens. The diagnostic specificity in acute leukemia is almost 100% when immunological markers are included *(71)*.

Cytogenetic diagnostics (karyotyping) of AML have lead to the development of therapeutic improvements, including guidance in the use of high risk therapeutic regimen in addition to molecular targeting therapeutics, for example, ATRA and imatinib mesylate (described below). However, more than half of patients will have no informative karyotype, and hope is that improved risk stratification of these patients will be obtained by the use of novel molecular methods *(5,15)*. A more detailed subclassification of AML can possibly be performed by proteomic profiling, but its role in future routine diagnostics is still unclear. However, work is in progress in this field which may lead to a better understanding of the protein network and protein specific diagnosis *(7,15)*.

4.1. Proteomics on Cell Extracts in AML Classification

Subgroups of leukemia patients can be profiled by comparing expression of protein biomarkers. As overviewed in Table 1, several comparative proteomics studies of leukemia have been published and several diagnostic biomarkers have been suggested. In an ideal research design, differential protein expression profiles in patient cells and in normal cell counterparts are compared. However, relevant normal myeloid progenitor cells may be difficult to isolate in amounts necessary to allow a proteomic comparison with leukemic blasts.

Kwak and coworkers *(50)* searched for serum biomarkers in 2DE protein profiles from 12 AML patients and healthy controls. Eight proteins predominantly expressed in the AML group were found (Table 1). In a study using total cell extract, 61 patient samples of acute leukemia blasts were analyzed by 2DE and MS identification *(7)* (Table 1). This resulted in identification of 23 differentially expressed proteins between the AML subgroups, of which seven proteins were highly expressed in AML M2 and/or AML M3, for example, human heparin-binding protein, and azurocidin. AML differentiation was correlated with elevated expression levels of myeloid-related proteins 8 and 14 and myeloperoxidase that is currently used in cytochemical diagnostics of AML *(72,73)*. In the same study, heat shock protein 27 and Op18 were highly expressed in ALL but not in AML. In a study comparing ALL and AML patients by 2DE, expression levels of the protein Op18 is most prominent in ALL *(52)* (Table 1). Op18 stimulates increased proliferation and downregulates differentiation by destabilizing the microtubules in vitro *(74)*. Antisense inhibition of Op18 expression in leukemia results in growth retardation, and Op18 has been proposed as a potential new target for anti-leukemic interventions *(52)*.

SELDI-based experiments have been performed to compare protein expression between three childhood leukemia cell lines (ALL cell lines 697 and REH, biphenotypic myelomonocytic cell line MV 4–11) and the Kasumi AML cell line *(75)*. A 8.3-kDa peak identified as

Table 1
Proteomics in Development of Biomarkers in Acute Leukemia

Leukemia	Upregulated proteins	Protein ID	Methodology	References
AML M3	NM23-H1	Gi29468184	2DE and MS	Cui et al. 2004 (7)
AML M2 and M3	Azurocidin	Gi28977		
	Human heparin-binding protein	Gi2981936		
	Myeloperoxidase	Gi77666942		
	Proteinase 3	GI1633225		
ALL	Cofilin 1	Gi5031635		
	Heat shock 27 kDA protein 1	Gigi4504517		
	Op18	Gi5031851		
	Aldehyde reductase	Gi1633300		
	Endoplasmic reticulum protein 29 precursor	Gi5803013		
	NM23-H1	Gi 9468184		
AML	Immunoglobulin heavy-chain variant	Gi 2443015	Patient sera are analyzed by blotting, 2DE and MS	Kwak et al. 2004 (50)
	Proteasome 26S ATPase	Gi2443015		
	Haptoglobin-1 (fragment)	Gi3337390		
AML with Flt3 mutation	Phosphorylated Stat3	P40763	Multiplexed flow cytometry	Irish et al. 2004 (15)

(Continued)

Table 1
(Continued)

Leukemia	Upregulated proteins	Protein ID	Methodology	References
ALL	Phosphorylated Stat5 Op18	P42229 / P51692 P16949	2DE, MS and Western blot	Hanash et al. 2002 (52)
ALL	Rho GDP dissociation inhibitor 2 Eno1 CAPZA1 protein γ-Actin	Gi10835002 Gi29792061 12652785 Gi178045	Patient sera are analyzed by blotting, 2DE, MS	Cui et al. 2005 (51)
Childhood ALL	C-terminal truncated ubiquitin		SELDI	Hegedus et al. 2005 (76)
AL with chromosomal abnormalities	NuMa	XP_006005	2DE and MS	Ota et al. 2003 (53)

2DE, two-dimensional polyacrylamide gel electrophoresis; ALL, acute lymphoblastic leukemia; AML, acute myeloid leukemia; MS, mass spectrometry; SELDI, surface-enhanced laser desorption/ionization.

Note that various material was used to determine upregulation of putative protein biomarkers. In (7), the comparison was performed within different French-American-British classification of AML, and AML versus ALL. The analysis in (50, 51, 76) compared sera from healthy individuals with sera from leukemic patients. A growth factor response analysis *in vitro* of 30 AML patients was used to determine the hyperresponding signalling nodes associated with Flt3 mutations (15). Hegedus and coworkers compared pretreatment bone marrow from children with ALL and AML (76). In (53), AC133+ haematopoietic stem cell fractions from bone marrow of 13 acute leukemia patients were compared.

C-terminal truncated ubiquitin (Ub) *(76)* was found to be a marker of the ALL cell lines (Table 1). Interestingly, as detected by SELDI technology, high cytosolic levels of Ub were recently reported to be associated with good prognosis in breast cancer *(77)*.

The identification of leukemia antigens that elicit an autoantibody response may have utility in cancer screening, both for diagnosis and prognosis. A 2DE approach has been applied to detect antigens that induce a humoral immune response *(51)*. Leukemia patients (16 AML and 5 ALL) were diagnosed according to the French-American-British classification, and samples were derived from bone marrow aspirates prior to chemotherapy. Sera from these patients in addition to 20 patients with solid tumors and 22 non-cancer controls were analyzed for autoantibody-based reactivity by incubating with blotted filters of proteins obtained from leukemia patients and separated on a 2DE gel. Among six autoantibody activities preferentially detected in the leukemia patients, an autoantibody against Rho GDP dissociation inhibitor 2 (Table 1) was found in sera from 71% of the leukemia patients compared to approximately 5% in patients with solid tumors and in healthy controls.

Stem cells from 13 patients of various leukemic subtypes have been isolated and screened with a 2DE approach to map differences in protein expression *(53)*. Eleven spots showed a significant difference in abundance between leukemia samples where NuMa, a nuclear protein that associates with the mitotic apparatus, was significantly related to the number of abnormal chromosomes in the leukemic blasts.

4.2. AML Stratification by Single-Cell Analysis of Signaling Responses

In a study by Irish and coworkers *(15)*, leukemia cell lines and primary AML cells were stimulated by growth factors and cytokines known to be involved in cell cycle regulation and cell death modulation. Granulocyte macrophage-colony stimulating factor (GM-CSF), granulocyte-colony stimulating factor (G-CSF), Flt3 ligand, interleukin-3 and interferon-γ act by activating their specific receptor, which is autophosphorylated and then leads to recruitment of adaptor proteins that propagate the signal by phosphorylation to the detected signal nodes. Activation of these signaling nodes reflects important pathways involved in chemoresistance. The phosphorylation level of Stat-1, Stat-3, Stat-5, Stat-6, p38 and Erk1/2 was determined in patient AML cells by single-cell flow cytometric analysis, generating cytokine-stimulated cancer network profiles *(15)*. Interestingly, this study illustrated that AML patients could be clustered in accordance to their pattern of potentiated signaling response to cytokines that correlated with genetics and disease outcome. This principle, examining the dynamics involved in signaling network profiles following cytokine stimulation, could be relevant to the revelation of diagnostic subgroups and novel therapeutic targets.

5. PHARMACOPROTEOMICS IN AML

Proteomics in drug discovery and development has been termed pharmacoproteomics. Even with the limitation of few pharmacoproteomic studies in leukemia, the examples below still serve to demonstrate that proteomics can be used to characterize the complex molecular mechanisms behind the clinical effects of current chemotherapy. Leukemic cells can relatively easily be sampled from patients for proteomics analysis in a time course analysis of therapy protocols. Recently, as a proof of principle, this was demonstrated in a study of p53 isoform signature changes during induction therapy of AML *(78)*. Both primary target proteins of therapeutics and the cascade pathways indirectly involved can be studied through proteomics. Consequently, biological mechanisms of drug toxicity and development of therapeutic resistance may be unveiled leading to the development of improved therapeutics in particular subclasses of AML.

5.1. All-Trans Retinoic Acid

Acute promyelocytic leukemia (APL) is associated with chromosomal translocations with a break point in the *RAR*α gene on chromosome 17 *(79)* most commonly fused with the *PML* gene on chromosome 15. The PML-RARα fusion protein acts as a dominant negative inhibitor of retinoid-dependent DNA transcription, which results in a differentiation arrest of the myeloid cells *(80)*. ATRA, a derivative of vitamin A, induces differentiation of APL cells to neutrophile granulocytes *(81)*. Treatment with ATRA in combination with anthracycline-based induction therapy and in postremission therapy leads to complete remission in at least 70% of APL patients *(82)*. ATRA binds to PML-RARα leading to degradation of the fusion protein and reactivation of target genes which signal diverse biological effects such as cell maturation, apoptosis and growth suppression *(83)*. Cellular effects of ATRA are studied to further reveal the regular mechanisms of inducing cell differentiation and silencing cell proliferation.

A number of proteomics studies show quantitative changes induced by ATRA therapy on leukemia cells (for review, see ref. *16*). These studies are based on methodology showing quantitative protein changes in overall proteins demanding relatively small sample amounts, in contrast to the in vitro studies on posttranslationally modified proteins as discussed below; still, most of these studies are performed on leukemia-derived cell lines instead of clinical samples *(16,84–88)*. A study on patient samples showed the protein BTG1 (B-cell translocation gene 1) to be induced in patients with complete remission after ATRA therapy, but not in non-remission patients *(54)*. Experimentally, this was supported by the observation that ATRA-induced differentiation of the HL-60 cell line highly induces the BTG1 level *(54)*.

5.2. Tyrosine Kinase Inhibitors

Approximately 100 different kinases in human cells phosphorylate proteins on tyrosine residues, regulating cell signaling with consequences on cellular functions like cell growth, proliferation, differentiation and apoptosis *(89)*. Mutational variants of several tyrosine kinases like the Abl, Platelet-Derived Growth Factor Receptor-(beta) (PDGFR-β), c-Kit, Flt-3 and others have been reported in leukemia *(90)*. Based on these observations, a wide range of small molecule kinase inhibitors has been developed to disrupt signaling pathways that promote cell proliferation and survival of cancer cells.

The kinase inhibitor imatinib mesylate (STI571, Gleevec) targets the cytoplasmic Abl tyrosine protein kinase, as well as the the Bcr-Abl fusion protein pathogenic for CML *(10,91–94)*. However, imatinib has so far proven limited effect in bcr-abl-positive ALL *(95)*, underscoring the hypothesis that multiple mutations and gene aberrations lay behind proliferation of acute leukemia blasts. Like most kinase inhibitors, imatinib has limited specificity and inhibits also c-Kit and the PDGFR in therapeutic concentrations. This is illustrated by the therapeutic benefit of imatinib in gastrointestinal stromal tumors with activating mutations of c-Kit or PDGFR-α *(96)*. The type of mutations of c-Kit or PDGFR predicts the therapeutic effect, and resistance develops mostly through secondary point mutations in the kinase genes. Likewise, imatinib is effective in the myeloproliferative disease hypereosinophilic syndrome, a disease that harbors a PDGFR gene rearrangement *(97)*. The c-Kit-positive AML patients in general do not respond to imatinib, whereas PDGF-β-positive AML patients may respond *(98,99)*, Even if imatinib demonstrates limited effect when used as a single agent in AML, it may be effective in combination with conventional therapeutics. However, selecting the AML patients who will respond on imatinib cotreatment will probably be crucial for its success. Determination of signal transduction responses in AML blasts may be a functional approach to identify pathways that may be targeted by kinase inhibitors *(15)*, and the first use of single-cell analysis of signaling nodes in AML patients in clinical trials may indicate its promising potential *(100)*.

Various experimental strategies have been used to determine the effect of imatinib and similar compounds on cellular signal transduction proteins. The effect of imatinib on the tyrosine phosphoproteome in bcr-abl-positive leukemia cell lines has been examined by immunoaffinity purification, multidimensional LC and MS *(60)* and resulted in identification of 64 sites of tyrosine phosphorylation corresponding to 32 different proteins. Among these, imatinib inhibited phosphorylation of several residues in Abl and Bcr.

Inhibitors belonging to the pyrido(2,3-d)pyrimidine class of compounds are effective against most of the imatinib-resistant bcr-abl variants isolated from CML. An immobilized pyrido(2,3-d)pyrimidine derivative was used to identify binding protein kinases through affinity column chromatography and MS-based identification of the eluted proteins *(101)*. Demonstrating the fact that pyrido(2,3-d)pyrimidine is a kinase inhibitor with low specificity, more than 30 human protein kinases were isolated and identified by the immobilized compound. A similar methodological approach was applied to study p38 kinase inhibitor SB203580 *(101,102)*, demonstrating that cyclin G-associated kinase and CK1 were almost as potently inhibited as p38α and RICK (Rip-like interacting caspase like apoptosis regulatory protein kinase) was even more sensitive to SB203580 than the original p38 target.

MS-based proteomics has been used to compare tyrosine-phosphorylated proteins in response to the receptor tyrosine kinase ligands Epidermal Growth Factor (EGF) and PDGF in differentation of human mesenchymal stem cells *(103)*. Differentiation of mesenchymal stem cells into bone-forming cells was stimulated by EGF but not PDGF; more than 90% of the tyrosine-phosphorylated signaling proteins were regulated by both ligands, whereas the phosphatidylinositol 3-kinase pathway was exclusively activated by PDGF. This methodological approach, using SILAC combined with modification-based affinity purification, could be relevant when mapping quantitative changes in protein phosphorylation in regulation of signaling networks in hematological stem cells and development of leukemia blasts (*104–106*).

About 30% of all human proteins can be modified by phosphorylation, while only 0.05% of the phosphorylated residues are tyrosine *(107)*. Therefore, detection levels and methodological sensitivity are limiting tyrosine kinase regulation studies. For all studies as mentioned here, cell lines have been used as models to study cellular mechanisms. Hopefully, with improved technologies, in even more ideal studies, researchers can measure effects on the phosphor proteome in patient samples during future clinical studies.

5.3. Histone Deacetylase Inhibitors

Lysine acetylation is another important posttranslational modification involved in regulation of protein activities in the human cell, suggested to be involved in leukemogenesis *(108)*. Acetyl groups on histone are removed by histone deacetylase (HDAC), resulting in a closed heterochromatin structure, which affects the accessibility of transcription factors to the DNA. Therapeutic HDAC inhibitors (HDACIs) upregulate acetylation of the chromatin structure thereby activating gene transcription and differentiation of cells *(109)*. Several myeloid transcription regulators regulate gene expression, at least in part, through their interactions with histone acetylase and deacetylase *(92)*. One example of particular relevance is the PML-RARα fusion gene product that binds to RAR target genes and then recruits HDAC complexes and transcription silencing *(80)*. A more frequent genetic abnormality in AML involves the genes encoding CBP/p300 histone acetylases, acting on histones and important intracellular regulators, that is, the antioncogene product p53.

Agents that inhibit HDACs are proposed to have therapeutic effects in AML and myelodysplasia *(110–113)*. Interestingly, leukemia patients expressing a second RARα fusion protein, PLZF-RARα, show poor response to ATRA therapy, whereas HDACIs exhibit anti-leukemic effects in patients expressing both types of RARα fusion proteins *(114)*. The HDACIs investigated in clinical studies of leukemia are butyrate derivatives, valproic acid and depsipeptide (DDP), and their limited but distinct effect has suggested further studies on HDACI in combination with other drugs *(108)*.

Methodologically relevant for future clinical studies, a LC-MS-based method has been developed to measure differential expression of histone posttranslational modifications prior to treatment with different HDACIs in several leukemia cell lines *(115)*. This study, which showed a general increase in acetylated forms of histone H4 after HDAC inhibition, also established an efficient and sensitive technique toward monitoring of histone modification in patients undergoing clinical treatment with HDACIs. In a recent study, the effects of trichostatin-A and DDP in mouse lymphosarcoma cells were studied, and acetylation pattern of histone H4 was established *(116)*. Flow cytometry represents another interesting technique for multiparameter analysis of histone acetylation to monitor HDACI efficacy *(117)*. This is a relevant biomarker study as the technique is established on patient samples in clinical trials.

Secondary effects of the HDACI butyrate have been tested in the human colon cancer cell lines HCT-116 and HT-29 utilizing 2DE and matrix-assisted laser desorption/ionization time-of-flight (MALDI-TOF) MS *(118)*. Various components of the Ub-proteasome system were altered, indicating that butyrate may regulate the level of key proteins involved in the control of cell cycle, apoptosis and differentiation. Protein expression subsequent to treatment with the HDACI trichostatin-A has been studied in pancreas ductual carcinoma cell lines using 2DE and MALDI-TOF MS *(119)*. Trichostatin-A appears to upregulate proteins which promote cell death and downregulate proteins that favor cell growth.

Valproic acid in combination with ATRA and theophyllamin is recently reported to attenuate phosphoprotein signaling nodes, determined by flow cytometric analysis of patient cells up to 10 days after start of therapy. Upregulation in phosphosignaling was accompanied by increased maturation of the cells and demonstrate the promising application of single-cell phosphoproteomic analysis as cancer cell biomarkers in development of new therapy *(100)*. However, this study describes one patient, and a more extended analysis of more patients is needed to determine if HDACIs may be used to indirectly target potentiated signaling in cancer cells.

5.4. Proteasome Inhibitors

Cellular protein degradation is implicating a highly conserved 76-aminoacid Ub polypeptide that is conjugated to lysine residues, targeting the protein for proteasomal destruction *(120)*. The initial Ub modification may be followed by chain formation through ligation of additional Ub molecules to the first, forming di-, tri-, tetra- or poly-Ub. The proteasome is involved in degradation of up to 80% of the proteins in a cell, including many of the key regulatory proteins involved in the cell cycle *(121)*. Inhibition of the proteasome has been established as a new therapeutic principle, involving anti-tumor mechanisms as growth inhibition and apoptosis induction through molecular mechanisms that are not completely understood *(122–127)*. Many actively proliferating malignant cells appear more sensitive to proteasome blockade compared to normal cells *(128)*. In some cancer cells, like in human leukemic cells, proteasomes have been reported to be abnormally highly expressed *(124)*, and studies have suggested that proteasome inhibitors may effectively target leukemic stem cells *(38)*.

Several classes of drugs have been developed to selectively inhibit the proteasome such as bortezomib *(129)*, peptide aldehydes (e.g., MG132) *(130)*, dipeptidyl boronic acids (e.g., PS-341) *(131)*, vinyl sulfone tripeptides (e.g., NLVS) *(132)*, and natural products for example, lactacystin, eponomycin and epoxomicin *(133)* as well as derivatives of epoxomicin, like PR-171 *(134)*. Epoxomicin was developed in an antitumor screen against murine B16 melanoma tumor *(135)*. Identification of the molecular targets of epoximicin was performed through MS/MS and Edman degradation from a cell extract of proteins after affinity chromatography with biotinylated epoximicin *(136)*.

The mechanisms responsible for the therapeutic efficacy of proteasome inhibition and of the mechanism behind sensitivity in certain cell types are unclear and remain to be identified. Methodological difficulties have been encountered in elucidation of the Ub system *(137)* as ubiquitinated forms of a specific protein are characterized by heterogeneous Molecular Weights (MW) and pI related to the ubiquitinated state (mono, multi and poly). Enrichment strategies and multidimensional chromatography, which separate independently of various MW and pI, should be chosen prior to MS-based identification of ubiquitinated proteins. Also, Ub-labeled proteins with short half-life and rapid turnover are often present in quantities too small to be detected by MS analysis *(138)*.

However, some studies have been performed on the ubiquitinated proteome according to treatment with proteasome inhibitors. Ubiquitinated proteins have been isolated from human breast cancer cell line by immunoaffinity purification after treatment with MG132 *(58)*. In another study, the general effect of the proteasome inhibitors, MG132 and lactacystin, on leukemia-derived cell lines has been analyzed by 2DE and MS *(139)*. A total of 39 protein spots were affected by proteasome inhibitors, including 11 new apoptosis-associated proteins. In particular, the accumulation of unmodified eIF-5A appeared to play important roles in apoptosis induced by these therapeutics *(139)*. Further studies should be performed in the future to elucidate biological consequences of treatment with novel proteasome inhibitors, that is, bortezomib and PR-171.

5.5. Drug Resistance

Another relevant field for pharmacoproteomic applications in AML research is elucidating the biological mechanisms of drug resistance. In drug resistant cells, alternative protein(s) forms are developed avoiding drug binding in active sites and/or executing signaling effects independent of the regulated native protein forms. Searching for drug resistance protein forms, proteomics seem a highly relevant methodology to pursue.

2DE analysis has been applied to resolve the mechanisms of vinca alkaloid-induced drug resistance in both ALL cell lines and NOD/SCID mouse xenograft models of ALL *(140,141)*. Not surprisingly, as the vinca alkaloids act on β-tubulin, several of the differentially expressed proteins in resistant ALL cells were associated with tubulin and/or actin cytoskeletons *(140)*. Proteomics studies comparing multidrug-resistant cell line HL-60/DOX and drug-sensitive HL-60 revealed proteins like protein disulfide isomerase precursor and proteasome α1 to be differentially expressed *(142)*.

Studying secondary effects of the HDAC inhibitor, butyrate, in the human colon cancer cell lines HCT-116 and HT-29 *(118)*, an upregulation of both antiapoptotic and proapoptotic proteins was observed in the HT-29 cell line, whereas in HCT-116 cells, the upregulation of proapoptotic proteins dominated. These results may explain the higher efficiency of butyrate treatment in HCT-116 compared to HT-29 *(118)* and consequently may be relevant to the appreciation of diverse effects in future AML trials.

5.6. Graft-Versus-Host Proteomics in Leukemia

Allogeneic hematopoietic stem cell transplantation (alloSCT) is an effective consolidation therapy in AML *(8)*, involving a T-cell graft-versus-leukemia effect that is important for persisting remission. However, a serious complication of alloSCT is when the donor T cells attack host tissues like skin, liver, stomach and/or intestine in a graft-versus-host disease. Potential use of proteomics in alloSCT is exemplified by an interesting study for identifying difficult manageable patients disposed to devastating graft-versus-host reactions *(143)*. In this clinical study, urine samples from 40 patients were collected at several time points after transplantation and peptides were separated and fractionated by capillary electrophoresis and HPLC. More than 1000 peptides were characterized and more than 1000 signal intensity (MS) spectra were collected for each sample, and software was specially developed to process the enormous amount of data generated. Polypeptide patterns excreted in the urine of patients were significantly different from those of healthy volunteers. No significant differences were detected comparing different conditioning regimens. Eighteen patients developed graft-versus-host disease after alloSCT. Sixteen differentially expressed polypeptides formed a pattern indicating graft-versus-host disease at an early stage. Sequencing the polypeptides allowed identification of leukotriene A4 hydrolase and serum albumin *(143)*. This study indicates that proteomics may allow diagnostics in easy collectible body fluids for complicated immunology-related diseases that need close follow-up.

6. CONCLUDING REMARKS

Proteomics studies are utilized to search for novel biomarkers to improve the diagnostic risk stratification of AML and to identify future therapeutic targets in more individualized therapy. This implicates design of therapy that is well tolerated in elderly patients. One major challenge in proteomics is its limitations in protein detection sensitivities. This is a major problem as many of the pivotal proteins for therapy response are suggested to be low abundant regulatory proteins. However, proteomic technologies are developing and sensitivities are increasing. The first reports on single-cell phosphoprotein examination indicate that this approach is useful in risk stratification as well as monitoring the effect of new therapy in AML patients. Ongoing and future proteomics studies will probably contribute to open new avenues of tailor-made molecular therapy, reducing today's limitations of treatment toxicity and efficiency.

ACKNOWLEDGMENTS

This study was supported by the Norwegian Research Council Functional Genomics Program (FUGE) grant number 151859, The Norwegian Cancer Society (Kreftforeningen) and Helse Vest.

CONFLICT OF INTEREST

None.

REFERENCES

1. Lowenberg, B.; Downing, J. R. and Burnett, A. (1999) N Engl J Med, 341, 1051–62.
2. Harris, N. L.; Jaffe, E. S.; Diebold, J.; Flandrin, G.; Muller-Hermelink, H. K.; Vardiman, J.; Lister, T. A. and Bloomfield, C. D. (1999) J Clin Oncol, 17, 3835–49.
3. Grimwade, D.; Walker, H.; Harrison, G.; Oliver, F.; Chatters, S.; Harrison, C. J.; Wheatley, K.; Burnett, A. K. and Goldstone, A. H. (2001) Blood, 98, 1312–20.
4. Bullinger, L.; Dohner, K.; Bair, E.; Frohling, S.; Schlenk, R. F.; Tibshirani, R.; Dohner, H. and Pollack, J. R. (2004) N Engl J Med, 350, 1605–16.
5. Valk, P. J.; Verhaak, R. G.; Beijen, M. A.; Erpelinck, C. A.; Barjesteh van Waalwijk van Doorn-Khosrovani, S.; Boer, J. M.; Beverloo, H. B.; Moorhouse, M. J.; van der Spek, P. J.; Lowenberg, B. and Delwel, R. (2004) N Engl J Med, 350, 1617–28.
6. Kohlmann, A.; Schoch, C.; Schnittger, S.; Dugas, M.; Hiddemann, W.; Kern, W. and Haferlach, T. (2004) Leukemia, 18, 63–71.
7. Cui, J. W.; Wang, J.; He, K.; Jin, B. F.; Wang, H. X.; Li, W.; Kang, L. H.; Hu, M. R.; Li, H. Y.; Yu, M.; Shen, B. F.; Wang, G. J. and Zhang, X. M. (2004) Clin Cancer Res, 10, 6887–96.
8. Cornelissen, J. J. and Lowenberg, B. (2005) Hematology (Am Soc Hematol Educ Program), 151–5.
9. Tallman, M. S. (2004) Semin Hematol, 41, 27–32.
10. Kantarjian, H.; Sawyers, C.; Hochhaus, A.; Guilhot, F.; Schiffer, C.; Gambacorti-Passerini, C.; Niederwieser, D.; Resta, D.; Capdeville, R.; Zoellner, U.; Talpaz, M.; Druker, B.; Goldman, J.; O'Brien, S. G.; Russell, N.; Fischer, T.; Ottmann, O.; Cony-Makhoul, P.; Facon, T.; Stone, R.; Miller, C.; Tallman, M.; Brown, R.; Schuster, M.; Loughran, T.; Gratwohl, A.; Mandelli, F.; Saglio, G.; Lazzarino, M.; Russo, D.; Baccarani, M. and Morra, E. (2002) N Engl J Med, 346, 645–52.
11. de Hoog, C. L. and Mann, M. (2004) Annu Rev Genomics Hum Genet, 5, 267–93.
12. Tyers, M. and Mann, M. (2003) Nature, 422, 193–7.
13. Cristea, I. M.; Gaskell, S. J. and Whetton, A. D. (2004) Blood, 103, 3624–34.
14. Gygi, S. P.; Rochon, Y.; Franza, B. R. and Aebersold, R. (1999) Mol Cell Biol, 19, 1720–30.
15. Irish, J. M.; Hovland, R.; Krutzik, P. O.; Perez, O. D.; Bruserud, O.; Gjertsen, B. T. and Nolan, G. P. (2004) Cell, 118, 217–28.
16. Sjoholt, G.; Anensen, N.; Wergeland, L.; Mc Cormack, E.; Bruserud, O. and Gjertsen, B. T. (2005) Curr Drug Target, 6, 631–46.
17. Gorg, A.; Weiss, W. and Dunn, M. J. (2004) Proteomics, 4, 3665–85.
18. O'Farrell, P. H. (1975) J Biol Chem, 250, 4007–21.
19. Shen, Y. and Smith, R. D. (2002) Electrophoresis, 23, 3106–24.
20. Wang, H. and Hanash, S. (2003) J Chromatogr B Analyt Technol Biomed Life Sci, 787, 11–8.
21. Cooper, J. W.; Wang, Y. and Lee, C. S. (2004) Electrophoresis, 25, 3913–26.
22. Binz, P. A.; Hochstrasser, D. F. and Appel, R. D. (2003) Clin Chem Lab Med, 41, 1540–51.
23. Aebersold, R. and Mann, M. (2003) Nature, 422, 198–207.
24. Sechi, S. and Oda, Y. (2003) Curr Opin Chem Biol, 7, 70–7.

25. Gygi, S. P.; Rist, B.; Gerber, S. A.; Turecek, F.; Gelb, M. H. and Aebersold, R. (1999) Nat Biotechnol, 17, 994–9.
26. DeSouza, L.; Diehl, G.; Rodrigues, M. J.; Guo, J.; Romaschin, A. D.; Colgan, T. J. and Siu, K. W. (2005) J Proteome Res, 4, 377–86.
27. Murphy, R. C. and Clay, K. L. (1982) Methods Enzymol, 86, 547–51.
28. Ong, S. E.; Blagoev, B.; Kratchmarova, I.; Kristensen, D. B.; Steen, H.; Pandey, A. and Mann, M. (2002) Mol Cell Proteomics, 1, 376–86.
29. Gerber, S. A.; Rush, J.; Stemman, O.; Kirschner, M. W. and Gygi, S. P. (2003) Proc Natl Acad Sci USA, 100, 6940–5.
30. Kirkpatrick, D. S.; Weldon, S. F.; Tsaprailis, G.; Liebler, D. C. and Gandolfi, A. J. (2005) Proteomics, 5, 2104–11.
31. Vermeulen, R.; Lan, Q.; Zhang, L.; Gunn, L.; McCarthy, D.; Woodbury, R. L.; McGuire, M.; Podust, V. N.; Li, G.; Chatterjee, N.; Mu, R.; Yin, S.; Rothman, N. and Smith, M. T. (2005) Proc Natl Acad Sci USA, 102, 17041–6.
32. Templin, M. F.; Stoll, D.; Schwenk, J. M.; Potz, O.; Kramer, S. and Joos, T. O. (2003) Proteomics, 3, 2155–66.
33. Seibert, V.; Wiesner, A.; Buschmann, T. and Meuer, J. (2004) Pathol Res Pract, 200, 83–94.
34. Diamond, D. L.; Zhang, Y.; Gaiger, A.; Smithgall, M.; Vedvick, T. S. and Carter, D. (2003) J Am Soc Mass Spectrom, 14, 760–5.
35. Williams, T. L.; Leopold, P. and Musser, S. (2002) Anal Chem, 74, 5807–13.
36. Gillette, M. A.; Mani, D. R. and Carr, S. A. (2005) J Proteome Res, 4, 1143–54.
37. Sachs, K.; Perez, O.; Pe'er, D.; Lauffenburger, D. A. and Nolan, G. P. (2005) Science, 308, 523–9.
38. Guzman, M. L.; Swiderski, C. F.; Howard, D. S.; Grimes, B. A.; Rossi, R. M.; Szilvassy, S. J. and Jordan, C. T. (2002) Proc Natl Acad Sci USA, 99, 16220–5.
39. Lash, G. E.; Scaife, P. J.; Innes, B. A.; Otun, H. A.; Robson, S. C.; Searle, R. F. and Bulmer, J. N. (2006) J Immunol Methods, 309, 205–8.
40. Vignali, D. A. (2000) J Immunol Methods, 243, 243–55.
41. Khan, I. H.; Mendoza, S.; Rhyne, P.; Ziman, M.; Tuscano, J.; Eisinger, D.; Kung, H.-J. and Luciw, P. A. (2006) Mol Cell Proteomics, 5, 758–68.
42. Rassenti, L. Z.; Huynh, L.; Toy, T. L.; Chen, L.; Keating, M. J.; Gribben, J. G.; Neuberg, D. S.; Flinn, I. W.; Rai, K. R.; Byrd, J. C.; Kay, N. E.; Greaves, A.; Weiss, A. and Kipps, T. J. (2004) N Engl J Med, 351, 893–901.
43. MacBeath, G. (2002) Nat Genet, 32 Suppl, 526–32.
44. Poetz, O.; Schwenk, J. M.; Kramer, S.; Stoll, D.; Templin, M. F. and Joos, T. O. (2005) Mech Ageing Dev, 126, 161–70.
45. Kovarova, H.; Hajduch, M.; Livingstone, M.; Dzubak, P. and Lefkovits, I. (2003) J Chromatogr B Analyt Technol Biomed Life Sci, 787, 53–61.
46. Kumble, K. D. (2003) Anal Bioanal Chem, 377, 812–9.
47. Espina, V.; Woodhouse, E. C.; Wulfkuhle, J.; Asmussen, H. D.; Petricoin, E. F., III and Liotta, L. A. (2004) J Immunol Methods, 290, 121–33.
48. Belov, L.; de la Vega, O.; dos Remedios, C. G.; Mulligan, S. P. and Christopherson, R. I. (2001) Cancer Res, 61, 4483–9.
49. Belov, L.; Huang, P.; Chrisp, J. S.; Mulligan, S. P. and Christopherson, R. I. (2005) J Immunol Methods, 305, 10–19.
50. Kwak, J. Y.; Ma, T. Z.; Yoo, M. J.; Choi, B. H.; Kim, H. G.; Kim, S. R.; Yim, C. Y. and Kwak, Y. G. (2004) Exp Hematol, 32, 836–42.

51. Cui, J. W.; Li, W. H.; Wang, J.; Li, A. L.; Li, H. Y.; Wang, H. X.; He, K.; Li, W.; Kang, L. H.; Yu, M.; Shen, B. F.; Wang, G. J. and Zhang, X. M. (2005) Mol Cell Proteomics, 4, 1718–24.
52. Hanash, S. M.; Madoz-Gurpide, J. and Misek, D. E. (2002) Leukemia, 16, 478–85.
53. Ota, J.; Yamashita, Y.; Okawa, K.; Kisanuki, H.; Fujiwara, S.; Ishikawa, M.; Lim Choi, Y.; Ueno, S.; Ohki, R.; Koinuma, K.; Wada, T.; Compton, D.; Kadoya, T. and Mano, H. (2003) Oncogene, 22, 5720–8.
54. Cho, J. W.; Kim, J. J.; Park, S. G.; Lee do, H.; Lee, S. C.; Kim, H. J.; Park, B. C. and Cho, S. (2004) Proteomics, 4, 3456–63.
55. Gjertsen, B. T.; Oyan, A. M.; Marzolf, B.; Hovland, R.; Gausdal, G.; Doskeland, S. O.; Dimitrov, K.; Golden, A.; Kalland, K. H.; Hood, L. and Bruserud, O. (2002) J Hematother Stem Cell Res, 11, 469–81.
56. Alaiya, A.; Al-Mohanna, M. and Linder, S. (2005) J Proteome Res, 4, 1213–22.
57. Jensen, O. N. (2004) Curr Opin Chem Biol, 8, 33–41.
58. Vasilescu, J.; Smith, J. C.; Ethier, M. and Figeys, D. (2005) J Proteome Res, 4, 2192–200.
59. Ballif, B. A.; Villen, J.; Beausoleil, S. A.; Schwartz, D. and Gygi, S. P. (2004) Mol Cell Proteomics, 3, 1093–101.
60. Salomon, A. R.; Ficarro, S. B.; Brill, L. M.; Brinker, A.; Phung, Q. T.; Ericson, C.; Sauer, K.; Brock, A.; Horn, D. M.; Schultz, P. G. and Peters, E. C. (2003) Proc Natl Acad Sci USA, 100, 443–8.
61. Steen, H.; Fernandez, M.; Ghaffari, S.; Pandey, A. and Mann, M. (2003) Mol Cell Proteomics, 2, 138–45.
62. Stasyk, T. and Huber, L. A. (2004) Proteomics, 4, 3704–16.
63. Anderson, L. (2005) J Physiol, 563, 23–60.
64. Jacobs, J. M.; Adkins, J. N.; Qian, W. J.; Liu, T.; Shen, Y.; Camp, D. G., II and Smith, R. D. (2005) J Proteome Res, 4, 1073–85.
65. Passegue, E.; Jamieson, C. H.; Ailles, L. E. and Weissman, I. L. (2003) Proc Natl Acad Sci USA, 100 Suppl 1, 11842–9.
66. Reya, T.; Morrison, S. J.; Clarke, M. F. and Weissman, I. L. (2001) Nature, 414, 105–11.
67. Akashi, K.; Traver, D.; Miyamoto, T. and Weissman, I. L. (2000) Nature, 404, 193–7.
68. Reuss-Borst, M. A.; Klein, G.; Waller, H. D. and Muller, C. A. (1995) Leukemia, 9, 869–74.
69. Sovalat, H.; Racadot, E.; Ojeda, M.; Lewandowski, H.; Chaboute, V. and Henon, P. (2003) J Hematother Stem Cell Res, 12, 473–89.
70. Matutes, E.; Pombo de Oliveira, M.; Foroni, L.; Morilla, R. and Catovsky, D. (1988) Br J Haematol, 69, 205–11.
71. Bain, B. and Catovsky, D. (1990) J Clin Pathol, 43, 882–7.
72. Tsuruta, T.; Tani, K.; Hoshika, A. and Asano, S. (1999) Leuk Lymphoma, 32, 257–67.
73. Bennett, J. M.; Catovsky, D.; Daniel, M. T.; Flandrin, G.; Galton, D. A.; Gralnick, H. R. and Sultan, C. (1976) Br J Haematol, 33, 451–8.
74. Belmont, L. D. and Mitchison, T. J. (1996) Cell, 84, 623–31.
75. Smith, M. T.; McHale, C. M.; Wiemels, J. L.; Zhang, L.; Wiencke, J. K.; Zheng, S.; Gunn, L.; Skibola, C. F.; Ma, X. and Buffler, P. A. (2005) Toxicol Appl Pharmacol, 206, 237–45.

76. Hegedus, C. M.; Gunn, L.; Skibola, C. F.; Zhang, L.; Shiao, R.; Fu, S.; Dalmasso, E. A.; Metayer, C.; Dahl, G. V.; Buffler, P. A. and Smith, M. T. (2005) Leukemia, 19, 1713–8.
77. Ricolleau, G.; Charbonnel, C.; Lode, L.; Loussouarn, D.; Joalland, M. P.; Bogumil, R.; Jourdain, S.; Minvielle, S.; Campone, M.; Deporte-Fety, R.; Campion, L. and Jezequel, P. (2006) Proteomics, 6, 1963–75.
78. Anensen, N., Oyan, A.M., Kalland, K.H., Bruserud, O., and Gjertsen, B.T. (2006) Clin Cancer Res., 12(13): 3985–92.
79. de The, H.; Chomienne, C.; Lanotte, M.; Degos, L. and Dejean, A. (1990) Nature, 347, 558–61.
80. Zhou, G. B.; Zhao, W. L.; Wang, Z. Y.; Chen, S. J. and Chen, Z. (2005) PLoS Med, 2, e12.
81. Huang, M. E.; Ye, Y. C.; Chen, S. R.; Chai, J. R.; Lu, J. X.; Zhoa, L.; Gu, L. J. and Wang, Z. Y. (1988) Blood, 72, 567–72.
82. Fenaux, P.; Chastang, C.; Chevret, S.; Sanz, M.; Dombret, H.; Archimbaud, E.; Fey, M.; Rayon, C.; Huguet, F.; Sotto, J. J.; Gardin, C.; Makhoul, P. C.; Travade, P.; Solary, E.; Fegueux, N.; Bordessoule, D.; Miguel, J. S.; Link, H.; Desablens, B.; Stamatoullas, A.; Deconinck, E.; Maloisel, F.; Castaigne, S.; Preudhomme, C. and Degos, L. (1999) Blood, 94, 1192–200.
83. Pitha-Rowe, I.; Petty, W. J.; Kitareewan, S. and Dmitrovsky, E. (2003) Leukemia, 17, 1723–30.
84. Guo, X.; Ying, W.; Wan, J.; Hu, Z.; Qian, X.; Zhang, H. and He, F. (2001) Electrophoresis, 22, 3067–75.
85. Harris, M. N.; Ozpolat, B.; Abdi, F.; Gu, S.; Legler, A.; Mawuenyega, K. G.; Tirado-Gomez, M.; Lopez-Berestein, G. and Chen, X. (2004) Blood, 104, 1314–23.
86. Wan, J.; Wang, J.; Cheng, H.; Yu, Y.; Xing, G.; Oiu, Z.; Qian, X. and He, F. (2001) Electrophoresis, 22, 3026–37.
87. Wang, D.; Jensen, R.; Gendeh, G.; Williams, K. and Pallavicini, M. G. (2004) J Proteome Res, 3, 627–35.
88. Lian, Z.; Wang, L.; Yamaga, S.; Bonds, W.; Beazer-Barclay, Y.; Kluger, Y.; Gerstein, M.; Newburger, P. E.; Berliner, N. and Weissman, S. M. (2001) Blood, 98, 513–24.
89. Cohen, P. (2002) Nat Rev Drug Discov, 1, 309–15.
90. Wadleigh, M.; Deangelo, D. J.; Griffin, J. D. and Stone, R. M. (2005) Blood, 105, 22–30.
91. Savage, D. G. and Antman, K. H. (2002) N Engl J Med, 346, 683–93.
92. Ravandi, F.; Kantarjian, H.; Giles, F. and Cortes, J. (2004) Cancer, 100, 441–54.
93. O'Brien, S. G.; Guilhot, F.; Larson, R. A.; Gathmann, I.; Baccarani, M.; Cervantes, F.; Cornelissen, J. J.; Fischer, T.; Hochhaus, A.; Hughes, T.; Lechner, K.; Nielsen, J. L.; Rousselot, P.; Reiffers, J.; Saglio, G.; Shepherd, J.; Simonsson, B.; Gratwohl, A.; Goldman, J. M.; Kantarjian, H.; Taylor, K.; Verhoef, G.; Bolton, A. E.; Capdeville, R. and Druker, B. J. (2003) N Engl J Med, 348, 994–1004.
94. Cortes, J.; Giles, F.; O'Brien, S.; Thomas, D.; Albitar, M.; Rios, M. B.; Talpaz, M.; Garcia-Manero, G.; Faderl, S.; Letvak, L.; Salvado, A. and Kantarjian, H. (2003) Cancer, 97, 2760–6.
95. Ottmann, O. G. and Wassmann, B. (2005) Hematology (Am Soc Hematol Educ Program), 118–22.

96. Shinomura, Y.; Kinoshita, K.; Tsutsui, S. and Hirota, S. (2005) J Gastroenterol, 40, 775–780.
97. Cortes, J.; Ault, P.; Koller, C.; Thomas, D.; Ferrajoli, A.; Wierda, W.; Rios, M. B.; Letvak, L.; Kaled, E. S. and Kantarjian, H. (2003) Blood, 101, 4714–6.
98. Kindler, T.; Breitenbuecher, F.; Marx, A.; Beck, J.; Hess, G.; Weinkauf, B.; Duyster, J.; Peschel, C.; Kirkpatrick, C. J.; Theobald, M.; Gschaidmeier, H.; Huber, C. and Fischer, T. (2004) Blood, 103, 3644–54.
99. Malagola, M.; Martinelli, G.; Rondoni, M.; Paolini, S.; Gaitani, S.; Arpinati, M.; Piccaluga, P. P.; Amabile, M.; Basi, C.; Ottaviani, E.; Candoni, A.; Gottardi, E.; Cilloni, D.; Bocchia, M.; Saglio, G.; Lauria, F.; Fanin, R.; Visani, G.; Marre, M. C.; Maderna, M.; Rancati, F.; Vinaccia, V.; Russo, D. and Baccarani, M. (2005) Blood, 105, 904; author reply 905.
100. Anensen, N.; Skavland, J.; Stapnes, C.; Ryningen, A.; Borresen-Dale, A. L.; Gjertsen, B. T. and Bruserud, O. (2006) Leukemia, 20, 734–6.
101. Wissing, J.; Godl, K.; Brehmer, D.; Blencke, S.; Weber, M.; Habenberger, P.; Stein-Gerlach, M.; Missio, A.; Cotten, M.; Muller, S. and Daub, H. (2004) Mol Cell Proteomics, 3, 1181–93.
102. Godl, K.; Wissing, J.; Kurtenbach, A.; Habenberger, P.; Blencke, S.; Gutbrod, H.; Salassidis, K.; Stein-Gerlach, M.; Missio, A.; Cotten, M. and Daub, H. (2003) Proc Natl Acad Sci USA, 100, 15434–9.
103. Kratchmarova, I.; Blagoev, B.; Haack-Sorensen, M.; Kassem, M. and Mann, M. (2005) Science, 308, 1472–7.
104. Blagoev, B.; Kratchmarova, I.; Ong, S. E.; Nielsen, M.; Foster, L. J. and Mann, M. (2003) Nat Biotechnol, 21, 315–8.
105. Blagoev, B.; Ong, S. E.; Kratchmarova, I. and Mann, M. (2004) Nat Biotechnol, 22, 1139–45.
106. Boeri Erba, E.; Bergatto, E.; Cabodi, S.; Silengo, L.; Tarone, G.; Defilippi, P. and Jensen, O. N. (2005) Mol Cell Proteomics, 4, 1107–21.
107. Hunter, T. (1998) Philos Trans R Soc Lond B Biol Sci, 353, 583–605.
108. Bruserud, O.; Stapnes, C.; Tronstad, K. J.; Ryningen, A.; Anensen, N. and Gjertsen, B. T. (2006) Expert Opin Ther Targets, 10, 51–68.
109. Melnick, A. and Licht, J. D. (2002) Curr Opin Hematol, 9, 322–32.
110. Phiel, C. J.; Zhang, F.; Huang, E. Y.; Guenther, M. G.; Lazar, M. A. and Klein, P. S. (2001) J Biol Chem, 276, 36734–41.
111. Gottlicher, M.; Minucci, S.; Zhu, P.; Kramer, O. H.; Schimpf, A.; Giavara, S.; Sleeman, J. P.; Lo Coco, F.; Nervi, C.; Pelicci, P. G. and Heinzel, T. (2001) EMBO J, 20, 6969–78.
112. Kuendgen, A.; Strupp, C.; Aivado, M.; Bernhardt, A.; Hildebrandt, B.; Haas, R.; Germing, U. and Gattermann, N. (2004) Blood, 104, 1266–9.
113. Insinga, A.; Monestiroli, S.; Ronzoni, S.; Gelmetti, V.; Marchesi, F.; Viale, A.; Altucci, L.; Nervi, C.; Minucci, S. and Pelicci, P. G. (2005) Nat Med, 11, 71–6.
114. Ferrara, F. F.; Fazi, F.; Bianchini, A.; Padula, F.; Gelmetti, V.; Minucci, S.; Mancini, M.; Pelicci, P. G.; Lo Coco, F. and Nervi, C. (2001) Cancer Res, 61, 2–7.
115. Zhang, L.; Freitas, M. A.; Wickham, J.; Parthun, M. R.; Klisovic, M. I.; Marcucci, G. and Byrd, J. C. (2004) J Am Soc Mass Spectrom, 15, 77–86.
116. Ren, C.; Zhang, L.; Freitas, M. A.; Ghoshal, K.; Parthun, M. R. and Jacob, S. T. (2005) J Am Soc Mass Spectrom, 16, 1641–53.
117. Ronzoni, S.; Faretta, M.; Ballarini, M.; Pelicci, P. and Minucci, S. (2005) Cytometry A, 66, 52–61.

118. Tan, S.; Seow, T. K.; Liang, R. C.; Koh, S.; Lee, C. P.; Chung, M. C. and Hooi, S. C. (2002) Int J Cancer, 98, 523–31.
119. Cecconi, D.; Scarpa, A.; Donadelli, M.; Palmieri, M.; Hamdan, M.; Astner, H. and Righetti, P. G. (2003) Electrophoresis, 24, 1871–8.
120. Schwartz, A. L. and Ciechanover, A. (1999) Annu Rev Med, 50, 57–74.
121. Smith, D. M. and Daniel, K. G. (2005) Lett Drug Des Discov, 2, 74–81.
122. Adams, J. (2004) Nat Rev Cancer, 4, 349–60.
123. Chauhan, D.; Hideshima, T. and Anderson, K. C. (2005) Annu Rev Pharmacol Toxicol, 45, 465–76.
124. Kumatori, A.; Tanaka, K.; Inamura, N.; Sone, S.; Ogura, T.; Matsumoto, T.; Tachikawa, T.; Shin, S. and Ichihara, A. (1990) Proc Natl Acad Sci USA, 87, 7071–5.
125. Pajonk, F. and McBride, W. H. (2001) Radiat Res, 156, 447–59.
126. Spano, J. P.; Bay, J. O.; Blay, J. Y. and Rixe, O. (2005) Bull Cancer, 92, E61–6, 945–52.
127. Bo Kim, K.; Fonseca, F. N. and Crews, C. M. (2005) Methods Enzymol, 399, 585–609.
128. Drexler, H. C. (1997) Proc Natl Acad Sci USA, 94, 855–60.
129. Rajkumar, S. V.; Richardson, P. G.; Hideshima, T. and Anderson, K. C. (2005) J Clin Oncol, 23, 630–9.
130. Lee, D. H. and Goldberg, A. L. (1998) Trends Cell Biol, 8, 397–403.
131. Gardner, R. C.; Assinder, S. J.; Christie, G.; Mason, G. G.; Markwell, R.; Wadsworth, H.; McLaughlin, M.; King, R.; Chabot-Fletcher, M. C.; Breton, J. J.; Allsop, D. and Rivett, A. J. (2000) Biochem J, 346 Pt 2, 447–54.
132. Adams, J.; Palombella, V. J. and Elliott, P. J. (2000) Invest New Drugs, 18, 109–21.
133. Sin, N.; Meng, L.; Auth, H. and Crews, C. M. (1998) Bioorg Med Chem, 6, 1209–17.
134. Kaiser, T., Kamal, H., Ran, K.A., Kolb, H.J., Holler, E., Ganser, A., Hertenstein, B., Mischak, H., Weissinger, E.M., Blood 2004; 104(2): 340–9.
135. Hanada, M.; Sugawara, K.; Kaneta, K.; Toda, S.; Nishiyama, Y.; Tomita, K.; Yamamoto, H.; Konishi, M. and Oki, T. (1992) J Antibiot (Tokyo), 45, 1746–52.
136. Meng, L.; Mohan, R.; Kwok, B. H.; Elofsson, M.; Sin, N. and Crews, C. M. (1999) Proc Natl Acad Sci USA, 96, 10403–8.
137. Kirkpatrick, D. S.; Denison, C. and Gygi, S. P. (2005) Nat Cell Biol, 7, 750–7.
138. Peng, J.; Schwartz, D.; Elias, J. E.; Thoreen, C. C.; Cheng, D.; Marsischky, G.; Roelofs, J.; Finley, D. and Gygi, S. P. (2003) Nat Biotechnol, 21, 921–6.
139. Jin, B. F.; He, K.; Wang, H. X.; Wang, J.; Zhou, T.; Lan, Y.; Hu, M. R.; Wei, K. H.; Yang, S. C.; Shen, B. F. and Zhang, X. M. (2003) Oncogene, 22, 4819–30.
140. Verrills, N. M.; Walsh, B. J.; Cobon, G. S.; Hains, P. G. and Kavallaris, M. (2003) J Biol Chem, 278, 45082–93.
141. Verrills, N. M.; Liem, N. L.; Liaw, T. Y.; Hood, B. D.; Lock, R. B. and Kavallaris, M. (2006) Proteomics, 6, 1681–94.
142. Chen, C. Y.; Jia, J. H.; Zhang, M. X.; Meng, Y. S.; Kong, D. X.; Pan, X. L. and Yu, X. P. (2005) Chin J Physiol, 48, 115–20.
143. Kaiser, T.; Kamal, H.; Rank, A.; Kolb, H. J.; Holler, E.; Ganser, A.; Hertenstein, B.; Mischak, H. and Weissinger, E. M. (2004) Blood, 104, 340–9.

9 New Tumor Biomarkers

Practical Considerations Prior to Clinical Application

Nils Brünner, Mads Holten-Andersen, Fred Sweep, John Foekens, Manfred Schmitt, Michael J. Duffy, on behalf of the EORTC PathoBiology Group

Contents

1. Introduction
2. Pre-Analytical, Analytical and Post-Analytical Aspects of Biomarker Assessment
3. Biomarkers for the Early Detection of Cancer
4. Prognostic Tumor Biomarkers
5. Predictive Tumor Biomarkers
6. Monitoring Tumor Biomarkers
7. Conclusion

Summary

This chapter describes the key elements in tumor biomarker development toward clinical use. In particular, it focuses on the analytical aspects of assay development and how to implement new tumor biomarkers in the clinical setting.

From: *Cancer Drug Discovery and Development
Cancer Proteomics: From Bench to Bedside*
Edited by: S. S. Daoud © Humana Press Inc., Totowa, NJ

Key Words: Pre-analytical variables; analytical variables; post-analytical variables; detection markers; screening markers; prognostic markers; predictive markers; monitoring markers

1. INTRODUCTION

Broadly defined, a tumour biomarker is a tool aiding the clinician to answer clinically relevant questions regarding cancerous disease *(1)*. However, a more specific definition characterizes a tumor biomarker as a molecule, a process or a substance that is altered quantitatively or qualitatively in pre-cancerous or cancerous conditions *(2)*. The alteration can be provoked by the tumor itself or by the surrounding stroma and is detectable by means of an established assay *(2)*. Thus, a tumor biomarker is a molecule produced by a tumor or by the body in response to a tumor and which aids cancer detection and/or management of a patient with cancer. The tumor biomarker can be protein, mRNA, DNA or a biologic process (e.g., apoptosis, angiogenesis and proliferation), and the detection assay can accordingly be of different formats ranging from complex gene arrays to immunohistochemical tests.

Regarding utility, tumor biomarkers can be divided into detection, diagnostic, prognostic, predictive and monitoring markers *(2)*. Obviously, a detection tumor biomarker is used for the detection of malignant disease in an individual and can be designated as a screening marker if used in larger asymptomatic populations. Preferably, such a detection biomarker should be tissue specific, be measurable in a readily available body fluid and be uninfluenced by benign diseases of the tissue/organ in question. Unfortunately, such an ideal marker does not exist for cancer.

A prognostic tumor biomarker enables the clinician to assess the risk of recurrence and subsequent death of the individual patient following intended curative surgery of the primary tumor. Hereby, it may be used to decide whether a patient should receive adjuvant therapy or can be spared of the adverse effects hereof. A therapy predictive biomarker will aid in estimating the likelihood of obtaining an objective response to a specific form of anti-cancer therapy and hereby aid in the selection of treatment strategies for the individual cancer patient. Finally, a monitoring tumor marker may be used in the post-operative surveillance of the cancer patient as it levels in blood may increase prior to the clinical detection of recurrent disease. However, to be of clinical value, the monitoring marker should alert the clinician at a time point when the recurrent cancer is still sensitive to available interventions.

An additional use of a monitoring marker is the evaluation of treatment response. The potential to use tumor biomarkers to determine treatment efficacy before information on objective response can be obtained opens up for the possibility to change or stop ongoing treatment if it shows to be ineffective. Such an approach will not only give the patient the option to receive second-line treatment at an earlier time point but will also have economical impact in saved expenses to otherwise ineffective treatment.

The field of tumor biomarkers has expanded widely since the identification of the Bence–Jones protein (immunoglobulin light chain) in urine from patients with multiple myeloma *(1)*. Current examples of tumor biomarkers in clinical use are manifold: prostate-specific antigen for early detection of prostate cancer in asymptomatic individuals, alpha-fetoprotein for the detection of tumors of the liver, testis or other germ cell line tumors, steroid hormone receptors (estrogen and progesterone receptors) for predicting response to treatment with endocrine therapy in breast cancer, immunostaining or Fluoresence In Situ Hybridization (FISH) for c-erbB2 for selecting for response to herceptin in breast cancer, Carcino Embryonic Antigen (CEA) for monitoring of colorectal cancer (CRC) and CA-125 for monitoring of ovarian cancer patients. Concurrently, with the identification of many new potential tumor biomarkers, the need for internationally accepted consensus guidelines for the establishment and validation of these new markers has become increasingly apparent. It is well known that strict guidelines are internationally defined and accepted for the introduction of new pharmacologic agents into the market. Similarly, introduction of new tumor biomarkers into the clinical management of cancer patients should be adhered to similar well-defined and internationally accepted conditions. Suggestions for such guidelines have been made by an expert panel convened by the American Society of Clinical Oncology (ASCO) defining potential specific uses of tumor biomarkers, specific requirements for the technical development as well as requirements for clinical trials to be fulfilled prior to implementation into the clinic *(2)*. The guidelines included a framework—the tumor marker utility grading system (TMUGS)—for the evaluation of already published tumor marker studies. By the use of the TMUGS, the available data on a potential tumor marker can be evaluated, and a certain level of evidence (LOE) can be assigned with which the marker can be considered for clinical use if found to be at a sufficiently LOE. These LOEs range from V to I, where V refers to small pilot studies, IV and III refer to retrospective studies of smaller (IV) or larger (III) size, II refers to prospective studies with marker testing as a secondary aim and I refers to either high-powered prospective studies with the primary aim of biological marker testing or meta-analyses of (several) studies at lower LOE stages of the marker in question. In general, most existing studies of tumor markers are of LOE IV or III.

In this respect, it is worthwhile to learn about the aims of the European Organization for Research and Treatment of Cancer (EORTC), Brussels, Belgium. The EORTC is a European Cancer Organization with the aim to develop, conduct, coordinate and stimulate laboratory and clinical cancer research to improve the management of cancer and related problems. This task can best be accomplished through the multidisciplinary, multinational efforts of clinical and basic research scientists and clinicians *(3)*. As the ultimate goal of the EORTC is to improve the standard of cancer treatment through the development of innovative drugs and more effective therapeutic strategies including surgery and radiotherapy, scientists and clinicians willing to participate in EORTC research are organized in clinical and laboratory groups. In cooperation with the clinical groups or task forces, the EORTC Laboratory Research Division, encompassing the Pharmacology and Molecular Mechanisms Group and the PathoBiology Group of the EORTC, focuses on pre-clinical testing of new drugs, functional imaging, pharmacology and pathology and on the assessment of tumor-associated biomarkers. Whereas the Pharmacology and Molecular Mechanisms Group centers on pharmacology, pharmacogenetics, pharmacogenomics, pharmacodynamics and the molecular mechanisms of anticancer drug effects/drug-related molecular pathology, the PathoBiology Group aims at reviewing tumor tissue specimens from cancer patients included in EORTC clinical trials, and as a translational aspect, conducting research and teaching in the life sciences focusing on detection, characterization, determination, and potential clinical application of tumor tissue markers and blood markers, associated with cancer disease progression and cancer metastasis. Thereby, attention is on biomarker development and implementation, accompanied by high-level quality assurance programs and the setting up of standard operating procedures for different pre-analytical, analytical, and post-analytical steps in biomarker development and implementation *(4,5)*.

The present review focuses on guidelines and practicalities with regard to development, validation and introduction of new tumor biomarkers for clinical management of cancer patients. In order to ease the reading, this chapter has been divided into sections dealing with pre-analytical, analytical and post-analytical aspects of biomarker assessment followed by chapters describing markers for early detection of cancer, prognostic evaluation, prediction of therapy response, postoperative surveillance and monitoring of therapy.

2. PRE-ANALYTICAL, ANALYTICAL AND POST-ANALYTICAL ASPECTS OF BIOMARKER ASSESSMENT

Although new biomarkers are being introduced for clinical decision making, a fundamental understanding of their clinical use and their underlying biological process is often lacking. The specifications of a test must address the clinical application. Remarkably, while biomarkers are often used in the clinical setting to provide additional information that will influence clinical decision making, only few guidelines (clinical and analytical) have been established to inform about how a biomarker should be clinically used for a certain type of cancer.

Biomarker assay results often are heterogeneous, depending on the composition of the specimen, method of specimen processing, and design and specificity of an assay. In addition, different statistical tests used to analyze the data may contribute to variation. It is clear that the development of common assay protocols, validation trials and proper quality control is highly needed before introduction of any new biological marker into the clinical routine.

Validation of an immuno(metric) assay method should include the following steps (see also ref. 6):

1. Selection of the appropriate test (based on literature data, experience, performance and availability);
2. Definition of performance requirements (specimen type, sample size, analytical range, precision, accuracy, sensitivity, specificity, recovery and interference); and
3. Statistical evaluation of data by state-of-the-art statistics.

In addition, the performance of an assay must be monitored by proper QA procedures aimed to achieve the desired high quality level of performance

2.1. Pre-analytical Aspects

Before a tumor specimen or blood sample enters the process of analyte quantification, several crucial steps have occurred outside the laboratory. The piece of tumor used for analysis should be as representative as possible of the whole tumor with regard to content of tumor cells, non-malignant cells, extracellular matrix, fat and necrotic spots. Because of the heterogeneity of a tumor (subclonal diversity), sampling bias may occur leading to different assay results if different areas of a tumor are analyzed. The size of a tissue specimen is thus important, and therefore, results from a tumor removed in toto by surgery may differ from those obtained assessing a fine needle aspirate. Transport of the tissue from the operation theatre to the laboratory should be done in a standard manner (on ice) and as quick as possible. Upon receipt of the tumor specimen by the laboratory, the material should be processed or snap frozen. Long term storage should be in low temperature-controlled containers. Disintegration/extraction of tissue samples should be according to internationally accepted protocols. Biomarkers may occur in different molecular forms, and the molecular forms present in the different subcellular fractions may vary between different tumors. This variation will even be greater when different extraction methods are employed, for example, including or excluding non-ionic detergents. A histological confirmation on presence of tumor tissue in the sample should always occur. Needless to say that error in this phase of evaluation is crucial and would lead to low integrity of data. Unfortunately, these steps fall beyond the evaluable QA processes performed in the laboratory (vide infra). Blood samples should be collected under standardized conditions (fasting, position, time of day and use of turnequet). Factors such as gender and age, menopausal status, race and so on should also be taken into consideration, as tumor marker levels may vary according to such biological variables. Preferably, plasma should be separate from blood cells immediately following blood withdrawal and frozen to avoid lysis of thrombocytes and/or biomarker degradation. Also, serum should be processed within a short time after collection. In this regard, centrifugation speed may influence tumor marker levels in the final sample preparation.

2.2. Analytical Aspects

An inherent problem of immuno(metric)assays is that different test kits may generate different test results, mainly because of differences in antibody specificity and/or affinity used in different test kits and because of use of different standards and reference materials provided with the kits. Prior to producing test results, a laboratory must verify or establish performance specifications for each method before its introduction to the laboratory. In particular, assays not yet in widespread use but showing promising clinical data must be validated carefully. The analytical specifications described in the following section should be assessed.

2.2.1. STANDARD/CALIBRATOR

Standards are used to relate the reading of an assay to the quantity of analyte to be measured. When a compound can be completely defined by physicochemical means, concentrations can be expressed in molar units. Protein analytes, however, may exist in different molecular forms, often exhibiting differences in their biological activities. Analytes extracted from tumor tissue or fluids may even be more different in nature from those present in the peripheral circulation of a patient. Assays should use well-defined and characterized standards, with as high purity as possible.

2.2.2. RECOVERY

When the standard of the assay is chemically different in nature from the analyte or when interactions of the analyte with the matrix or other compounds are involved, different biomarker concentration may be observed when the tumor specimen is analyzed at different dilutions. The addition of a fixed amount of known standard provides information on the identity of the analyte vs. standard and/or on any interfering process.

2.2.3. ACCURACY

The accuracy of an assay is defined as the agreement between the best estimate of a quantity and its true value. Only for analytes for which a reference method is available are such comparison is possible. For most biomarkers, however, the true value does not exist, as no reference method has been established.

2.2.4. PRECISION

The precision of an assay is defined as the agreement between replicate measurements. The precision of a biomarker determination varies depending on whether duplicate determinations are performed in one sample, different samples in the same batch or in different batches and so on. For validation of an assay, the intra-sample, intra-assay precision performance is the minimum required. A thumb rule is that between assays Coefficient of Variation (CVs) below 10% is acceptable.

2.2.5. SENSITIVITY

The limit of detection is defined by the lowest concentration detected that is significantly different from zero (also called analytical sensitivity). The limit of quantification is the lowest concentration at which a test result can be reliably measured with a CV of <20% (also called functional sensitivity). Laboratories should not report below the functional sensitivity limit. As lower amounts of tumor tissues are becoming available, the need for higher sensitivity assays is required. Defining such low thresholds requires a high degree of reproducibility of assay results.

2.2.6. SPECIFICITY

The specificity of immunoassays strongly depends on antibody characteristics. Polyclonal or monoclonal antibodies or mixtures of both are applied in different test kits. In general, polyclonal antibodies have higher sensitivity but also increased chance that one of these antibodies will recognize an epitope on a different antigen, resulting in cross-reactivity and thereby decreasing specificity. In contrast, monoclonal antibodies have increased specificity. In order to avoid false-positive test results, all assays should be checked for cross-reacting compounds.

2.2.7. LINEARITY

Some assay procedures demand that samples are diluted to within a specified range of protein content prior to the assay. Values multiplied by the dilution factor should give the same results, irrespective of extent of dilution. Such experiments are often referred to as parallelism studies because dilution of samples should parallel the standard curve. Linearity studies are used to assess the working range of an assay.

2.2.8. INTERFERENCES

Blood of patients may contain heterophilic antibodies that may interfere in sandwich assays leading to false-positive or false-negative test results. One example is induction of human anti-mouse *(7)*. Assays designed for measurement of biomarkers in tumor tissue cytosols or tumor tissue extracts should not be used to test plasma or serum samples without checking whether these assays have been adequately tested (see above) in these sample types.

The above-mentioned assay characteristics should be assessed by the investigator when a new assay is introduced into the laboratory. It has to be considered, however, that there are day-to-day, performer-to-performer and batch-to-batch differences in assay performance. For daily consistency, use of controls for testing between assay precision and participation in external quality assurance programs is mandatory.

2.3. Post-Analytical Aspects

After an assay has been performed, the read-outs should be processed and interpolated in the standard curves. It has been shown that recalculation of raw data of a participant of an international QC trial by a common computer program reduced the between-laboratory variation *(8)*. It is preferable that Standard Statistical Procedures are in place before the assay results are available.

2.4. Quality Assessment

Within Europe, a multitude of translational multi-center cancer studies have been coordinated by the EORTC. Within this consortium, the PathoBiology Group was established to research and advice on common, or equivalent, methodologies for tumor biomarker assays and to ensure that appropriate external quality assessment (EQA) schemes are applied to all laboratories measuring biomarkers in biopsies from patients entering EORTC-conducted studies/trials. For the past 20 years, large-scale EQA trials have been carried out organized by the central QA laboratory of the PathoBiology Group *(9–11)*.

Defined protocols for internal and external QA should be part of routine practice in the laboratory, and new assays should be thoroughly validated upon first use. Every assay consists of a measurement procedure to determine analyte levels of a certain biomarker and a control procedure in which by measurement of control samples the validity of the measurement of the samples can be checked. Control samples, and comparison of its values against control limits, should always be an integral part of the assay procedure. For internal QC (IQC) purposes, the laboratory must include samples of different concentrations of control material. This opens the way for multirule/decision control procedures *(12)*. Repeated measurement of control samples allows determination of the imprecision of the assay system. In addition to the use of IQC for day-to-day assay monitoring, the long-term trend in assay performance should be regularly checked in order to detect any shift or drift. For external QC purposes, preparations distributed by a reference laboratory should be included. These EQA programs serve to monitor long-term assay performance within a laboratory. Moreover, they provide comparison of assay results between laboratories. Based on our EORTC QA studies, enduring inter-laboratory consistencies were achieved after agreement was reached on the best method of assay and as soon as external QA was available *(13)*.

3. BIOMARKERS FOR THE EARLY DETECTION OF CANCER

For many cancer types, it is believed that early detection increases the chance of cure. Thus, requirement for a screening test is that it detects early stages and thereby potentially curable disease. An alternative to the development of new treatment modalities is thus the development of test methods for early detection of cancer.

The optimal tumor detection marker is one that can detect the cancer disease in a population of asymptomatic individuals. Such a tumor biological marker is named a screening marker. The challenges of detection or screening tests lie in obtaining a high specificity, high sensitivity and high compliance. Receiver operating characteristics curves are often used to graphically demonstrate the relationships between specificity and sensitivity of a particular tumor biological marker. The curve is produced based on tumor marker values obtained from diseased as well as non-diseased individuals. Specificity of a screening test describes the number of non-diseased individuals who are correctly defined as non-diseased. Sensitivity describes the percentage of diseased individuals who are correctly identified by the test. Compliance describes the percentage of individuals who are willing to take the test. The level of specificity required for a particular test very much depends on the subsequent procedures needed. For example, a screening marker for breast cancer could be considered as a "pre-mammography" test, and the requirements for specificity would be low. This is only true if the sensitivity at the same time was approaching 100%. Thus, in the screened positive population, all the breast cancer patients would be present (sensitivity of 100%), while in the screened negative population (the size of this population depends on the specificity), the women would avoid mammography. If the test was a simple blood test, a proportion of women who would otherwise be subjected to the expensive and time-consuming mammography with its discomfort and potential risk of irradiation-induced cancers would not be in need of the mammography.

One example is prostate-specific antigen (PSA), which is a serological tumor biological marker that is being used to screen asymptomatic men for the presence of prostate cancer. An elevated PSA level results in subsequent biopsy of the prostate to verify the diagnosis of cancer. This test has a rather low specificity (many false-positive cases), while the sensitivity is not accurately defined due to lack of studies in which all PSA-tested individuals are biopsied.

Another example of a screening test is the fecal occult blood testing (FOBT) in which stool is being analyzed for the presence of hemoglobin. The FOBT has clearly shown an increase in number of patients diagnosed with early stage CRC and that early detection translates into increased survival of screened patients *(14–16)*. However, the FOBT is unacceptable for many individuals and therefore the compliance to FOBT is relatively low (approximately 30–50%). When calculating "clinical sensitivity" which is sensitivity multiplied with compliance, the FOBT is severely hampered by very low values (20–30%) for this important end point.

4. PROGNOSTIC TUMOR BIOMARKERS

Prognostic factors or markers provide information related to the risk of disease relapse or death, independent of therapy *(17)*. Strictly defined, prognostic factors predict the natural history of patients who receive no adjuvant systemic therapy. Prognostic factors are mostly used to justify the administration of adjuvant therapy. Although these factors may also be used to withhold therapy (i.e., if they indicate risk of relapse is minimal), they are rarely used for such purposes *(17)*.

However in early stage cancer patients, it may be relevant to gain information on the individual patients' risk of recurrence and subsequent disease-related death. Such information can be used to select early stage "high-risk" patients for systemic anti-cancer therapy, while early stage "low-risk" patients may be spared, the unnecessary side effects of a treatment that is not needed. Furthermore, the use of prognostic markers in this patient category would also have a major impact on the economic health system in savings of expenses related to otherwise unnecessary treatment. Two good examples are derived from breast cancer and from colon cancer, respectively. In breast cancer, the majority of stage I patients are categorized as "high-risk" patients with a recurrence rate of approximately 30% over 5 years. These patients are offered systemic treatment knowing that 70% of these are already cured by the primary surgery. This is the reason for an intense scientific activity aiming at identifying new and robust prognostic factors for this patient category. Especially transcriptional profiling has added a novel dimension to cancer diagnostics and prognosis, identifying molecular markers (gene signatures) capable of differentiating tumors and disease outcome beyond the discriminatory power of histopathologic evaluation and established prognostic factors *(18)*. In colon cancer, which is a disease that mainly affects the older population, it is known that patients with stage II disease have a 5-year risk of recurrence and subsequent death of approximately 30%. In contrast to stage I breast cancer patients, where the recurrence numbers are the same as for stage II colon cancer, the colon cancer patients are not offered systemic therapy. A prognostic marker, that would identify the "high-risk" patients among the stage II colon cancer patients, would allow for focused therapy of this patient group *(19)*. It is also noteworthy that in future clinical trials with adjuvant treatment of early stage cancer patients then inclusion of a prognostic marker would allow selection of "high-risk" patients for the study. Such an approach would be due to increased percentage of events lower than the number of patients needed in the study (patients already cured by the primary surgery are excluded) and also lower than the post-treatment observation time.

Although a large number of putative prognostic factors have been described for cancer (for review, see refs *20,21*), few have been clinically validated. Amongst the best-validated prognostic biomarkers are urokinase plasminogen activator (uPA) and Plasminogen Activator Inhibitor-1 (PAI-1) in breast cancer. Based on both a randomized prospective trial *(22)* and a pooled analysis *(23)*, both these factors were shown to be independent predictors of outcome in patients with lymph node-negative breast cancer.

Axillary node-negative patients with low levels of uPA and PAI-1 have a low probability of developing recurrent disease and thus may be candidates for being able to avoid the side effects and costs of adjuvant chemotherapy *(22,23)*. On the other hand, axillary node-negative patients with high levels of uPA and/or PAI-1 have a relatively high risk of progression, that is, approximately similar to women with three positive lymph nodes *(24)*. Consequently, these patients should receive adjuvant chemotherapy, especially as patients with increased levels of uPA and PAI-1 show an enhanced benefit from adjuvant chemotherapy *(23,24)*.

In evaluation new prognostic biomarkers, it is important to avoid bias in patient/sample selection. Consequently, prognostic biomarkers are best validated in a prospective study. Provided bias can be eliminated; samples from retrospective collections, however, may also provide reliable data. The advantage of using prospectively collected samples is that patients follow-up information may be available, thus shortening the length of the study.

Irrespective of whether a retrospective or prospective evaluation is performed, validation of results is essential. Different types of validation have been described *(25)*. Initially, internal validation should be performed. This may be performed either by splitting the data into two groups with one group used for training the model and the other for testing the model or by a form of cross-validation that is based on repeated model development and testing on random data partitions *(26)*. Prior to clinical use, independent validation in an external institute is necessary *(25)*.

5. PREDICTIVE TUMOR BIOMARKERS

The first therapy predictive markers to be described were the estrogen and progesterone receptors. By measuring these receptors in breast cancer tissue, patients with receptor-positive tumors can be selected for endocrine therapy, while patients with receptor-negative tumors can be offered another treatment modality because the likelihood of responding to endocrine treatment in a receptor-negative breast cancer patients is close to zero. Recently, with the appearance of targeted therapy, new predictive tests have been developed and validated in clinical studies. For example, measurement of Human Epidermal Growth Factor-2 (HER2) in breast cancer tissue is being used to guide the physicians when selecting breast cancer patients for herceptin treatment, and measurements of c-kit and Platelet Derived Growth Factor receptor (PDGFr) are being used to select patients for treatment with Glivec *(27)*.

It is likely that future types of targeted therapy will be accompanied with assay platforms to determine the presence of the target in question and thereby be of significant helping in future management of cancer patients. Thus, it is clear that predictive molecular assays should be established before initiation of clinical trials for new targeted anticancer agents if the specificity and usefulness of these drugs are to be meaningfully evaluated in the population of patients most likely to benefit from the treatment.

At present, there is no test in routine use to determine whether a cancer patient will benefit from conventional chemotherapy in its broad sense or to individual types of conventional chemotherapy. If such a test existed, it would allow for a tailor-made approach resulting in individualized treatment. This would also imply that for those patients having resistant tumors, such an approach would not only spare a large number of these patients from side effects induced by ineffective chemotherapy, but would also have major impact on the economic health system in savings of expenses related to otherwise ineffective treatment. The field of cancer drug discovery clearly needs to turn greater attention to the problem of identifying responsive/resistant subsets of patients early in the development process and needs to utilize the knowledge obtained through molecular and cellular studies of cancer biology.

When validating new predictive markers for clinical use, the pivotal clinical trial will have to be randomized and compare marker-guided treatment against unguided treatment and then use objective response rate, time to progression and overall survival as end points. The EORTC is at present planning such trials. In addition, these trials include prospective standard operative procedure-guided tissue and blood sampling.

6. MONITORING TUMOR BIOMARKERS

6.1. Postoperative Surveillance

Following curative resection for primary cancer, it is now a common practice to follow-up patients at regular intervals. The main aim of this surveillance is to detect recurrent/metastatic disease at an early stage, the assumption being that the early detection of disease progression followed by the initiation of therapy enhances patient outcome compared to initiating therapy when the patient is symptomatic.

For most cancers, however, there is little evidence that intensive follow-up improves prognosis. An exception, however, is CEA for the surveillance of patients who have undergone curative resection for CRC. Four independent meta-analyses have compared outcome in CRC patients undergoing intensive follow-up (with regular CEA determinations) versus those with minimal or no regular follow-up *(28–31)*. All these meta-analyses concluded that the use of an intensive follow-up regime resulted in a modest but statistically significant better outcome than either no or minimal follow-up *(28–31)*. Another malignancy in which follow-up with serial determinations of tumor markers is mandatory is in non-seminomatous germ cell tumors. In this situation, regular measurement of Alfa Fetp-Protein (AFP) and Human Chorionic Gonadotrophin (HCG) is now standard practice *(32)*.

In order to use a tumor marker in surveillance, it is first necessary to define the extent of elevation in marker levels that constitutes a significant increase. Over the years, various empirical definitions of tumor marker increases have been proposed including a doubling in marker level or increases of 25–30%. According to Tuxen et al. *(33)*, this alteration should be based on both the analytical variation of the assay used and the normal background intra-individual biological variation of the marker. Thus, in order for a change in marker concentrations to be significant ($p < 0.05$), the change must exceed random fluctuations due to both analytical and biological variations. Based on this definition, it has been calculated that successive PSA levels must change by at least 50% to be significant *(34)*.

Having established the magnitude of alteration that constitutes a significant increase in marker concentration, it is necessary to clinically validate this increment, that is, show that by using this increase, one can detect recurrent/metastatic disease earlier than with the use of standard clinical or radiological criteria. Assuming that serial determinations of marker provides a lead-time vis-à-vis standard procedures, the next step is to demonstrate that the initiation of therapy based on this tumor marker lead-time enhances patient disease-free survival, overall survival, quality of life or reduced costs of care. Ideally, the answer to this question should be addressed in a prospective randomized trial. Such a trial would randomize asymptomatic subjects into two groups as follows. One group would undergo treatment following a confirmed marker increase while the other would start therapy when recurrence is detected using standard criteria. An example of such a trial is currently underway investigating the potential value of CA 125 for the early detection of recurrent ovarian cancer *(35)*.

6.2. Monitoring Therapy

Most anti-cancer therapies are effective in only a proportion of treated patients, and most patients undergoing systemic therapy suffer from toxic side effects of the treatment. It is therefore important to know as quickly as possible if patients are benefiting from the administered therapy. Although there are exceptions, most patients responding to treatment show decreasing levels of markers while those with progressive disease usually exhibit increasing levels. As with markers in surveillance, only a small number of markers have been validated for monitoring therapy in advanced cancer.

One of the best examples is CA 125 for monitoring treatment in patients with ovarian cancer. Based on initial pilot studies, Rustin et al. *(36)* proposed the following definitions for determining response of ovarian cancer to chemotherapy. Response according to CA 125 occurred if there was either a 50% or a 75% reduction in CA 125 levels *(37)*. These definitions have been retrospectively tested in 19 phase 2 clinical trials investigating 14 different cytotoxic drugs for recurrent ovarian cancer *(36–38)*. Overall, responses based on CA 125 were similar to those based on standard criteria, leading Rustin and colleagues *(36–38)* to suggest that the 50 and 75% response criteria could substitute for standard responses in phase 2 clinical trials evaluating new treatments for ovarian cancer. As assessable disease is found in only a minority of patients with cancer of the ovary following debulking surgery, use of these definitions would increase the number of patients participating in clinical trials *(35)*.

Currently, for determining response to therapy in advanced cancer, criteria such as Union International Contra Cancer (UICC) or Response Evaluation Criteria in Solid Tumors (RECIST) *(39)* are used. If tumor markers are to be used, they must be at least as accurate as these criteria. As with surveillance, a prospective randomized trial is necessary in order to validate a marker for monitoring therapy in patients with advanced malignancy. Such a trial would randomize patients with advanced disease into two groups as follows. In one group, disease progression would be defined by a confirmed increase in marker level and in the other based on standard methods. End points should include quality of life, cost of care and overall survival.

7. CONCLUSION

New biomarkers should not be included in large clinical studies unless the assay methods are carefully evaluated and common assay protocols, common standards and reference preparations allowing proper EQA are available. Although considerable progress has been made, forces must be combined to assure that best technology is applied in the evaluation of biomarkers. Only the stringent application of QA systems enables a consistent assessment of the clinical value of biomarkers. Also, any introduction of biomarkers in the clinical management of cancer patients should be preceded by a prospective clinical study that unequivocally demonstrates the clinical benefit of using the marker. Such clinical studies can either be performed in conjunction with a therapy study or as a study with the primary goal of validating the biomarker. The EORTC-PathoBiology Group, the National Cancer Institute, Bethesda, MD, USA and the ASCO are organizing annual international meetings where most of the above-mentioned aspects are being discussed.

REFERENCES

1. Diamandis EP. Tumor markers: past, present, and future. In Tumor Markers: Physiology, Pathobiology, Technology, and Clinical Applications, EP Diamandis, H Fritsche, Jr, H Lilja, D Chan, M Schwartz (eds). Washington, DC, American Association for Clinical Chemistry (AACC) Press, 2002, pp 3–8.
2. Hayes DF, Bast RC, Desch CE, Fritsche H, Jr, Kemeny NE, Jessup JM, Locker GY, Macdonald JS, Mennel RG, Norton L, Ravdin P, Taube S, Winn RJ. Tumor marker utility grading system: a framework to evaluate clinical utility of tumor markers. J Natl Cancer Inst 1996;88:1456–1466.
3. Meunier, F. European Organisation for Research and Treatment of Cancer. Brussels, EORTC, 2006 (http://www.eortc.be).
4. Schmitt M, Harbeck N, Daidone MG, Brunner N, Duffy MJ, Foekens JA, Sweep FC. Identification, validation, and clinical implementation of tumor-associated biomarkers to improve therapy concepts, survival, and quality of life of cancer patients: tasks of the Receptor and Biomarker Group of the European Organization for Research and Treatment of Cancer. Int J Oncol 2004;5: 1397–1406.
5. Schrohl AS, Holten-Andersen M, Sweep F, Schmitt M, Harbeck N, Foekens J, Brunner N. European Organisation for Research and Treatment of Cancer (EORTC) Receptor and Biomarker Group. Tumor markers: from laboratory to clinical utility. Mol Cell Proteomics 2003;6:378–87.
6. Westgard JO, Barry PL. Cost-Effective Quality Control: Managing the Quality and Productivity of Analytical Processes. Washington, DC, American Association for Clinical Chemistry (AACC) Press, 1995.
7. Moseley KR, Knapp RC, Haisma HJ. An assay for the detection of human anti-murine immunoglobin in the presence of CA 125 antigen. J Immunol Methods 1988;106:1–6.
8. Koenders A, Thorpe SM, on behalf of the EORTC Receptor Group Standardization of steroid receptor assays in human breast cancer-I. Reproducibility of oestradiol and progesterone receptor assays. Eur J Cancer Clin Oncol 1983;19:1221–1229.
9. Geurts-Moespot J, Leake R, Benraad TJ, Sweep CGJ. Twenty years of experience with the steroid receptor External Quality Assessment program - the paradigm for tumour biomarker EQA studies (review). Int J Oncol 2000;17:13–22.
10. Sweep CGJ, Geurts-Moespot J. EORTC external quality assurance program for ER and PgR measurements: trial 1998/1999. Int J Biol Markers 2000;15:62–69.
11. Sweep CGJ, Geurts-Moespot J, Grebenschikov N, de Witte JH, Heuvel JJTM, Schmitt M, Duffy MJ, Jänicke F, Kramer MD, Foekens JA, Brünner N, Brugal G, Pedersen AN, Benraad TJ. External quality assessment of trans-European multicentre antigen determinations (ELISA) of urokinase-type plasminogen activator (uPA) and its type-1 inhibitor (PAI-1) in human breast cancer tissue extracts. Br J Cancer 1998;78:1434–1441.
12. Westgard JO, Barry PL, Hunt MR, Groth T. A multi-rule Shewart chart for quality control in clinical chemistry. Clin Chem 1981;27:493–501.
13. Koenders A, Benraad TJ. Standardization of steroid receptor analysis in breast cancer biopsies: EORTC receptor group. Recent Results Cancer Res 1984;91: 129–135.
14. Hardcastle JD, Chamberlain JO, Robinson MH, Moss SM, Amar SS, Balfour TW, James PD, Mangham CM. Randomised controlled trial of faecal-occult-blood screening for colorectal cancer [see comments]. Lancet 1996;348:1472–1477.

15. Kronborg O, Fenger C, Olsen J, Jorgensen OD, Sondergaard O. Randomised study of screening for colorectal cancer with faecal-occult-blood test [see comments]. Lancet 1996;348:1467–1471.
16. Selby JV, Friedman GD, Quesenberry CP, Jr, Weiss NS. A case-control study of screening sigmoidoscopy and mortality from colorectal cancer. N Engl J Med 1992;326:653–657.
17. Henderson IC, Patek AJ. The relationship between prognostic and predictive factors in the management of breast cancer. Breast Cancer Res Treat 1998;52: 261–288.
18. Basil CF, Zhao Y, Zavaglia K, Jin P, Panelli MC, Voiculescu S, Mandruzzato S, Lee HM, Seliger B, Freedman RS, Taylor PR, Hu N, Zanovello P, Marincola FM, Wang E. Common cancer biomarkers. Cancer Res 2006;66:2953–2961.
19. Holten-Andersen MN, Stephens RW, Nielsen HJ, Murphy G, Christensen IJ, Stetler-Stevenson W, Brunner N. High preoperative plasma tissue inhibitor of metalloproteinase-1 levels are associated with short survival of patients with colorectal cancer. Clin Cancer Res 2000;6:4292–4299.
20. Duffy MJ. Clinical uses of tumor markers: a critical review. Crit Rev Clin Lab Sci 2001;38:225–262.
21. Thomas CMG, Sweep CGJ. Serum tumor markers: past, state of the art and future. Int J Biol Markers 2001;16:73–86.
22. Janicke F, Prechtl A, Thomssen C, et al. For the German Chemo No Study Group. Randomized adjuvant chemotherapy trial in high-risk node-negative breast cancer patients identified by urokinase-type plasminogen activator and plasminogen activator inhibitor type 1. J Natl Cancer Inst 2001;93:913–920.
23. Look M P, van Putten WLJ, Duffy MJ, et al. Pooled analysis of prognostic impact of tumor biological factors uPA and PAI-1 in 8377 breast cancer patients. J Natl Cancer Inst 2002;94:116–128.
24. Harbeck N, Kates RE, Schmitt M. Clinical relevance of invasion factors urokinase-type plasminogen activator and plasminogen activator inhibitor type 1 for individualised therapy in primary breast cancer is greatest when used in combination. J Clin Oncol 2002;20:1000–1007.
25. Harbeck N, Kates RE, Look MP, et al. Enhanced benefit from adjuvant chemotherapy in breast cancer patients classified high-risk according to urokinase-type plasminogen activator (uPA) and plasminogen activator inhibitor type 1 (N = 3424). Cancer Res 2002;62:4617–4622.
26. Simon R. Development and validation of therapeutically relevant multi-gene biomarker classifiers. J Natl Cancer Inst 2005;97:866–867.
27. Duffy MJ. Predictive markers in breast and other cancers. Clin Chem 2005;51: 494–503.
28. Bruinvels DJ, Stiggelbout AM, Kievit J, van Houwelingen HC, Habbema DF, van de Velde CH. Follow-up of colorectal cancer: a meta-analysis. Ann Surg 1994;219:174–182.
29. Rosen M, Chan L, Beart RW, Vukasin P, Anthone G. Follow-up of colorectal cancer: a meta analysis. Dis Colon Rectum 1998;41:1116–1126.
30. Renehan AG, Egger M, Saunders MP, O'Dwyer ST. Impact on survival of intensive follow up after curative resection for colorectal cancer: systematic review and meta-analysis of randomised trials. BMJ 2002;324:813–816.
31. Figueredo A, Rumble RB, Maroun J, et al. Follow-up of patients with curatively resected colorectal cancer. BMC Cancer 2003;3:26–39.
32. Bosl GJ, Motzer RJ. Testicular germ-cell cancer. N Engl J Med 1977;337:242–251.

33. Tuxen MK, Soletormos G, Rustin GJS, et al. Biological variation and analytical imprecision of CA 125 in patients with ovarian cancer. Scand J Clin Lab Invest 2000;69:713–722.
34. Soletormos G, Semjonow A, Sibley PEC, et al. Biological variation of total prostate-specific antigen: a survey of published estimates and consequence for clinical practice. Clin Chem 2005;51:1342–1351.
35. Bridgewater JA, Nelstrop AE, Rustn GJS, et al. Comparison of standard and of CA 125 response criteria in patients with epithelial ovarian cancer treated with platinum or paclitaxel, J Clin Oncol 2003;17:501–508.
36. Rustin GJS, Nelstrop AE, McClean P, et al. Defining response of ovarian carcinoma to initial chemotherapy according to serum CA 125. J Clin Oncol 1996;14:1545–1551.
37. Rustin GJS. Use of CA 125 to assess response to new agents in ovarian cancer trials. J Clin Oncol 2003;21:187s–193s.
38. Rustin GJS, Nelstrop AE, Bebtzen SM, et al. Selection of active drugs for ovarian cancer based on CA 125 and standard response rates in phase II trials. J Clin Oncol 2000;18:1733–1739.
39. Therasse P, Arbuck SG, Eisenhauer EA, et al. New guidelines to evaluate response to treatment in solid tumors. J Natl Cancer Inst 2000; 92:205–216.

IV Bioinformatics and Regulatory Aspects of Proteomics

10 Annotating the Human Proteome

From Establishing a Parts List to a Tool for Target Identification

Rolf Apweiler and Michael Mueller

CONTENTS

1. INTRODUCTION
2. DEFINING THE PROTEOME IN HEALTH AND DISEASE
3. THE NEED FOR STANDARDS: DATA DISSEMINATION, EXCHANGE AND ACCESS
4. CONNECTING THE DATA: THE UNIPROT KNOWLEDGEBASE AND ITS HUMAN PROTEOMICS INITIATIVE, AND INTEGR8
5. CONCLUSION

SUMMARY

The completion of the human genome has shifted the attention from deciphering the sequence to the identification and characterization of the functional components including genes. Improved gene prediction algorithms together with existing transcript and protein information have enabled the identification of most exons in a genome. Availability of the 'parts list' has enabled systematic interrogation of gene function on genome, transcriptome and

From: *Cancer Drug Discovery and Development
Cancer Proteomics: From Bench to Bedside*
Edited by: S. S. Daoud © Humana Press Inc., Totowa, NJ

proteome level. Methodological and technological advancements particularly in mass spectrometry-based proteomics have fostered the development of tools to determine the proteomic signature of various human cells in health and disease, through the construction of comprehensive protein expression and protein modification profiles as well as protein interaction and protein localization maps. Studying gene function on protein level is vital to the understanding of molecular mechanisms underlying pathological conditions including cancer as variations in protein isoforms and protein quantity underlying a disease phenotype can often not be deduced from sequence or transcript level genomics alone. Besides providing a greater understanding of carcinogenesis, proteomics also holds the promise of identifying new targets for diagnostics, prognosis and therapeutics.

Key Words: Human, proteomics, functional annotation, data integration, data standardization

1. INTRODUCTION

Twenty years ago, Renato Dulbecco viewed the sequencing of the human genome as 'a turning point in cancer research' *(1)*. However, any genome is only as valuable as its annotation *(2)*, and the immense task of adding value to the human genome sequence through identification of its functional elements including protein coding genes and elucidation of their function has only just begun. When the first draft of the human genome sequence was published in 2001 *(3,4)*, the number of protein coding genes was estimated to be around 30,000–40,000. With the availability of the near finished sequence of the human euchromatic genome *(5)*, along with expanded cDNA libraries, genomes of other organisms and improved computational methods for gene prediction, this number has been adjusted to 20,000–25,000 protein coding genes. Although this number is lower than initially anticipated, the classical reductionist approach to function determination is no longer sufficient. With the availability of the genome sequence an array of 'omics' technologies to probe the function of many or all genes and their products in a single experiment has emerged. The field of proteomics has produced tools to systematically study proteins including their identification in complex protein mixtures, measure changes in their abundance, distribution, and modifications, study protein–protein interactions, determine protein structure and subcellular localization as well as characterize their biochemical activity. Each approach individually produces valuable information; however, eventually only an integrated analysis of data from different

Chapter 10 / Annotating the Human Proteome

'omics' domains will permit a system level understanding of how cells, tissues and organisms perform their function and what the changes underlying pathological conditions including cancer are. To this end, it is important to establish standards for data capturing, storage and exchange to allow straightforward access and integration of available information (Fig. 1).

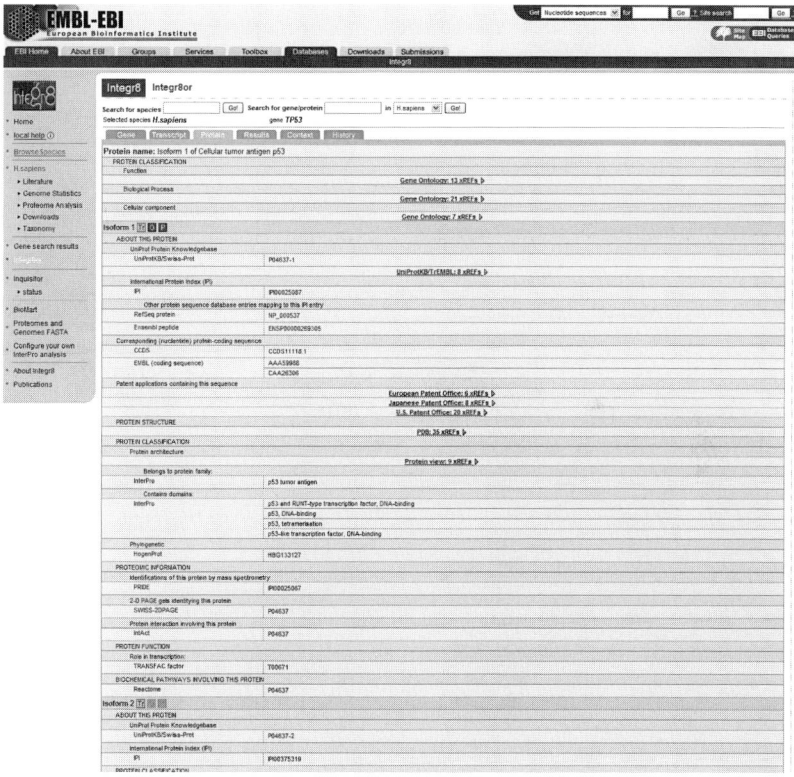

Fig. 1. From the parts list to the system. Information on diverse aspects of protein function under normal and pathological conditions is generated using a wide range of proteomics technologies. Data from different proteomics domains are gathered in public repositories. Data standards assure seamless data import, export and exchange; minimum reporting requirements for experiments enable a meaningful data (re-)analysis by third parties; and the use of ontologies and controlled vocabularies facilitates data comparison and integration, a prerequisite for studies on the system level.

This chapter discusses current approaches to interrogate protein function as well as efforts to coordinate and standardize proteomics data generation, capturing, exchange and integration.

2. DEFINING THE PROTEOME IN HEALTH AND DISEASE

The relatively low number of human genes suggests that complexity of human biology is achieved through regulation on the transcriptional, post-transcriptional and post-translational level. Alternative splicing and translation *(6)* and post-translational modification (PTM), such as phosphorylation, glycosylation and proteolytic cleavage, result in a 'proteomic stratification' generating a protein population of a size several magnitudes larger than the number of genes encoding them. It has been estimated that the human proteome comprises up to 1,000,000 different protein species *(7)*. Annotating the 'normal' human proteome and detecting alterations correlating with pathogenesis that could potentially serve as biomarkers for diseases, thus, seems to be a daunting task. However, through the rapid advancements of available technologies as well as coordination of global proteomics efforts, this goal might be achieved earlier than anticipated.

2.1. Splice Variants

Alternative splicing modulates subcellular localization, and activity of proteins through insertion or deletion of functional domains, subcellular sorting signals or transmembrane regions *(8–10)* has been implicated in a wide range of physiological and pathophysiological processes. Recent studies suggest that between 42 and 74% of human multi-exon genes are alternatively spliced *(10–12)*, and it has been estimated that 15% of point mutations that cause human genetic disease affect splicing *(13)*. Full-length cDNA sequencing projects provide gold-standard definitions of transcripts and are making substantial progress towards characterization of splice isoforms. The H-Invitational database (H-InvDB) *(14)* integrates human transcript information from six high-throughput cDNA sequencing projects comprising annotation of 56,419 full-length cDNAs grouped into 25,585 cDNA clusters. Sequencing based approaches are labour-intense and expensive. A technology more amenable to high-throughput analysis are exon junction arrays that have been used successfully to monitor splicing events of approximately 10000 multi-exon genes across diverse tissues *(12)*. Alternative splicing events derived from computational approaches are provided by the Alternative Splicing Database (ASD) *(15)*. The database also contains information curated from the literature.

2.2. Post-Translational Modification

PTM of proteins plays a critical role in the regulation of biological processes by affecting various aspects of protein behaviour such as activity, turnover, localization and molecular interactions. The number of documented protein co-translational modification and PTM has now exceeded 400. The most common modifications include phosphorylation, glycosylation, methylation, acetylation, nitration, sulphation, lipidation, ubiquitination and proteolytic cleavage. Large-scale studies of PTM are currently focused on phosphorylation and glycosylation.

2.2.1. PHOSPHORYLATION

Phosphorylation is a reversible PTM involved in the regulation of enzyme activity, signalling cascades and modulation of molecular interactions. Deregulation of phosphorylation in signalling pathways controlling cell proliferation and apoptosis is a feature of many types of cancers, and drugs targeting these pathways represent promising therapeutic tools.

There are a number of experimental approaches to map phosphorylation in protein mixtures. Methods based on two-dimensional (2D) gel electrophoresis detect phosphorylated proteins by anti-phospho-amino-acid antibodies, metabolic labelling or phosphatase treatment. Protein spots are then identified by mass spectrometry. Alternatively phosphorylated proteins can be identified in complex mixtures by LC MS/MS *(16,17)* either with or without prior affinity purification. In the latter case, fragmentation spectra that could not be assigned to proteins in a database of unmodified sequences are used in a second search iteration allowing for modification. Another approach to map phosphorylation sites is the proteolysis of proteins by phospho-specific cleavage, followed by mass spectrometry and deduction of phosphorylation sites from the cleavage pattern *(18)*. Protein microarrays are an emerging technology that will play an important role in phosphorylation profiling as well *(19,20)*; however, prior knowledge of targets is required here. More recently, quantitative mass spectrometry methods have been incorporated into phosphoproteomics *(21–24)*.

PhosphoSite *(25)* and Phospho.ELM *(26)* are two bioinformatics resources that provide curated information on in vivo phosphorylation sites of human and mouse proteins while the protein kinase resource *(27)* and the kinase pathway database *(28)* provide annotation of at least 518 putative protein kinases encoded in the human genome *(29)*.

2.2.2. GLYCOSYLATION

Glycosylation PTM through transfer of glycans or carbohydrates to proteins is a complex process requiring the concerted action of a series of glycosyl transferases each catalyzing a specific step in the pathway. Evolution of this complicated process suggests important functions of glycoproteins which are, however, poorly understood. Nearly half of all proteins are potentially glycosylated *(30)*, and oligosaccharides are implicated in the regulation of protein folding, Endoplasmic Reticulum (ER) to Golgi transport, cell–cell communication and immune response. There is also growing evidence for the implication of glycans in tumor growth and metastasis *(31–34)*.

Due to the non-template-driven nature of glycosylation, a direct study of function is difficult as it is not possible to knock out specific structures to evaluate their effect on the phenotype. A number of different technologies are employed to study glycosylation including mass spectrometry *(35–38)*, Nuclear Magnetic Resonance (NMR) *(39)* and liquid chromatography for glycan 'sequencing' *(40,41)*. Glycan microarrays are used for glycan–protein interaction profiling *(42)*, and new techniques to fluorescently label glycans have enabled quantification of glycan species on the array *(43)* while the carbohydrate-binding specificity of lectins is exploited to measure protein glycosylation states using lectin microarrays *(44–46)*.

Resources providing databases and bioinformatics tools for glycobiology include the KEGG Glycan database *(47)* and http://www.glycosciences.de hosted by the German Cancer Research Centre. Several standards for the representation of glycan structures have emerged to facilitate exchange and comparison of glycomics data *(48–50)*.

2.2.3. PROTEOLYTIC PROCESSING

Proteolytic cleavage is an important non-eversible PTM controlling the fate of proteins by influencing their subcellular localization and activity. Proteolysis can also give rise to proteins, so-called cryptic neoproteins with functions different from the parent protein it was derived from. It has become evident that proteolysis is a highly selective and regulated process playing a role in cellular processes going beyond catabolism, including DNA replication, cell proliferation, cell cycle, differentiation, migration and apoptosis *(51–53)*. Given the functional relevance of proteolysis, it is not surprising that a mis-regulation of protease activity underlies several pathological conditions including cancer *(52,53)*. The human genome encodes 561 proteases and protease homologues *(54)* on which information is available from the MEROPS database *(55)*. The resource provides summaries on each protease and its inhibitors and is cross-referenced with UniProt Knowledgebase (UniProtKB). As it is not possible to predict peptidase cleavage sites computationally at present because substrate structure plays an important role, and thus, these sites have to be determined experimentally. Several proteomics studies based on isotope-coded affinity tag (ICAT) labelling or 2D gel electrophoresis followed by MS/MS, and yeast two-hybrid screens have been dedicated to the discovery of new protease targets and have lead to the identification, for example, of new potential metalloproteinase *(56,57)* and caspase substrates *(58)*.

2.3. Subcellular Localization

Eukaryotic cells and especially mammalian cells are highly compartmentalized. The function of a protein, therefore, is often strongly correlated with its localization, and it has been shown that aberrations in protein localization influence tumour progression *(59)*. Development of methods to systematically green fluoresce protein (GFP) tag proteins using full-length cDNA libraries *(60)* in combination with microscope-based high-throughput visual screening *(61)* has enabled analysis of protein localization on a genome-wide scale. In mammalian cells, a large-scale analysis of protein localization is complicated by splice variants *(10)* and the wealth of different cell types. Thus, resolving the human 'localizome' is currently limited to only a fraction of proteins encoded in the human genome. As part of a large-scale project to systematically analyse protein function, the German Cancer Research Centre in collaboration with EMBL Heidelberg has determined the localization of approximately 1000 proteins encoded by full-length cDNAs generated by the German cDNA Consortium focusing on novel proteins with completely unknown function. Data are made public through the LIFEdb database *(62)* and the GFP-cDNA localization project (http://gfp-cdna.embl.de).

Classical subcellular fractionation methods from cell biology, followed by 2D gel electrophoresis and mass spectrometry, are applied

in organelle proteomics to catalogue the protein constituents of cellular compartments and components. Several proteomics
studies have analysed human organelles and other cellular structures including nucleus *(63,64)*, nuclear membrane *(65)*, GR-Golgi intermediate compartment (ERGIC) *(66)*, Golgi *(67)*, exosomes *(68,69)*, centrosome *(70)*, mitotic spindle *(71)* and mitochondria *(72)*.

Organelle DB *(73)* is a database of protein localization compiling information on known protein constituents of organelles as well as subcellular structures and protein complexes including manually curated localization data from several subcellular proteomic studies. Currently, the database contains localization information for more then 3900 human genes. Other resources are dedicated to specific organelle proteomes such as MitoRes *(74)* and Mitomap *(75)*, databases of mitochondrial proteins, the Nuclear Protein Database *(76)* and NMP-db of nuclear proteins *(77)* or the Hera database *(78)* of human proteins located in the endoplasmic reticulum.

2.4. Tissue Expression

While comprehensive information of tissue expression is available on the transcript level *(79)*, establishment of a similar resource for protein expression proves to be much more challenging due to the requirement of unique binding reagents specific to individual proteins. Large-scale projects are now underway to raise antibodies against each and every human protein with the aim to interrogate tissue expression on the protein level *(80,81)*. The Swedish Human Protein Atlas (HPA) *(82)* has been set up to explore the human proteome using antibody-based proteomics. Antibodies against human proteins are systematically raised and then used in high-throughput immunohistochemistry screens *(83)* on tissue microarrays *(84)* for expression profiling in normal and cancer tissue. Images are electronically captured and annotated.

Bioinformatics is employed to select epitopes used as antigens for antibody selection taking into account information on protein families and domains *(85)*. HPA generates so-called monospecific antibodies (msAbs) *(86)* through immunization of rabbits and subsequent affinity purification of the polyclonal immunoglobulins from immune sera. msAbs are potentially more versatile then monoclonal antibodies as the mixture of different antibodies recognizing the denatured as well as native form of the protein enables their use in immunohistochemistry, Western blots and pull down experiments *(86)*. Currently, expression information for more than 700 proteins in 48 different normal tissue types and 20 different types of cancer is accessible through the HPR web page (http://www.proteinatlas.org).

2.5. Abundance

Changes in gene expression involved in tumorigenesis have been studied extensively on mRNA level in recent years. However, the collective phenotypic endpoint of mechanisms regulating gene expression is protein abundance, which due to post-transcriptional mechanisms to regulate gene expression such as control of translation rate and mRNA stability cannot be deduced from mRNA expression data *(87)*. Methods to quantify protein expression are based on 2D gel electrophoresis, mass spectrometry or protein arrays. Quantification by conventional 2D gel electrophoresis involves matching of equivalent spots on gels used to separate protein samples from different experimental conditions, determination of differentially expressed proteins by comparing spot intensities and their identification by mass spectrometry. Two-dimensional fluorescence-difference gel electrophoresis (DIGE) *(88)* allows running two samples on the same gel by labelling them with different fluorescent dyes in vitro. This permits an accurate overlay of the two samples increasing the confidence and sensitivity with which differential expression is detected by overcoming the problem of inter-gel variability. Furthermore, the dynamic range of DIGE of 10^{-4} enables the detection of proteins expressed at relatively low levels.

A disadvantage of gel-based method is the bias against membrane proteins and low abundance proteins. Gel-free methods for protein expression profiling involve differential labelling of sample and control with stable isotopes. The relative ratio of heavy and light peptides derived from tryptic digest of labelled protein samples is then measured by mass spectrometry. Labelling can be performed in vitro after protein purification by chemical derivatization as well as in vivo by metabolic labelling. The prototype of expression profiling based on differential labelling is ICAT-based quantification *(89)*, which is based on in vitro tagging of purified proteins with a reagent consisting of a reactive group directed against cysteine, a polyether linker region labelled with deuterium, and biotin. Pooled and digested samples are then subjected to cation and avidin chromatography to reduce sample complexity followed by quantification of $^2H/^1H$ ratios by mass spectrometry. A range of other methods targeting different functional groups of the polypeptide chain or using different labels have been developed *(90)*.

The above approaches represent methods to measure relative protein abundance between two or more samples. Absolute protein quantification can be achieved by introducing isotopically labelled peptide standards of known quantity into a sample, usually during or after protein digestion. Application of this strategy on a global level has been limited by the requirement to synthesize suitable peptide standards on a larger scale. Recently, a method was introduced to produce peptide standards by de novo design of a gene encoding an artificial protein that is a concatemer of tryptic peptides *(91)*. The stable isotope is incorporated metabolically during heterologous expression of the protein.

A relatively recent approach to measure protein abundance is based on protein microarrays. There are two types of abundance-based microarrays. The first type, capture arrays, is produced by spotting analyte-specific capture molecules, for example, antibodies or affibodies on the array surface *(92)*. The sample is then applied onto the chip for target detection. The second type, reverse phase protein blots, is created by spotting the samples themselves on a slide and then probe with the analyte-specific reagent *(93,94)*. Analytes are detected either by direct labelling of the sample with a detectable marker or by using a labelled analyte-specific reagent. Protein microarrays have been used for protein profiling in cancer research *(94–97)*.

2.6. Protein–Protein Interaction

Identification of potential interaction partners of proteins involved in cancer will help understand their biological role in the context of functional modules and pathways and ultimately help to discover novel, therapeutic targets. As protein interaction mapping reveals relationships between proteins, it can also help to understand the effects a therapeutic intervention will have on cellular processes and the overall function of a cell. While genome-scale interaction is available for several model organisms since some time *(98–101)*, large-scale human interaction data sets have become available only recently *(102,103)*, although covering only a subset of the proteome. Besides these large-scale data sets, previous studies have investigated interactions in the context of specific biological processes such as tumor necrosis factor (TNF)-alpha/NF-kappa B signal transduction *(104)*, the Hsp90 interactome *(105)*, EGF signaling *(106)* or the SMAD signalling pathway *(107)*. All but the later study which is based on yeast two-hybrid have used affinity tagging and purification to detect complex components. Interaction information obtained by this approach is less error prone but not suitable to detect transient interactions likely to play an important role in signalling pathways *(108)*. Recently, a luminescence-based method has been employed to monitor transient interactions under physiological conditions *(109)* overcoming the drawback of yeast two-hybrid screens detecting interactions in an artificial milieu.

A number of databases capturing information on protein–protein interaction have been established in recent years. Resources curating human interaction data include IntAct *(110)*, BIND *(111)*, MINT *(112)*, DIP *(113)* and BioGrid *(114)*.

3. THE NEED FOR STANDARDS: DATA DISSEMINATION, EXCHANGE AND ACCESS

The success of genomics was made possible by large-scale collaborative projects sequencing the same clones many times and sharing the data by releasing the data in standardized form into public databases. In this way, the genome and cDNA projects led not only to scientific breakthroughs like deciphering genomes and the transcript landscape but also created community resources. These community resources enabled other scientists to harvest already in an early stage the fruits of the investments in these projects too.

Proteomics is currently still dominated by single-lab contributions; most experiments, especially the large-scale ones, are due to its costs rarely repeated, and the data end in various formats in various places.

To realize the power of proteomics, it is necessary to learn from genomics. Money in large-scale proteomics will be only worth spending if the return on investment will lead to community resources such as large data collections usable by and of interest for all life scientists or large scientific communities like the cancer community. To achieve this, many technical obstacles need tackling, especially because the rate of production of proteomics data continues to increase as high-throughput approaches become commonplace; the complexity of these data sets is also increasing as both workflows and instrumentation evolve. This creates two linked problems; to handle this volume of data would in itself present logistical challenges, but that these data are both voluminous and complex mandates sophisticated data handling techniques. A corollary of the commercial origin of much of the technology deployed in proteomics is that the data are not just complex and voluminous, they are frequently in proprietary formats; computational access to these is often contingent on the availability of a particular vendor's software. To enable the harvesting of proteomics data directly at the source of production in a common format is a prerequisite towards systematic data capturing, an area where proteomics still lags significantly behind.

Guidelines on what to report as part of a proteomics publication are being developed *(115)*, and formats on how to represent the data in a standardized manner are emerging *(116,117)*. Other efforts in establishing data standardization largely centre on the increasing use of controlled vocabularies and ontologies to standardize the way metadata and annotations are provided in data files. The most established and widely used ontology is probably the Gene Ontology (GO) *(118)* to describe attributes of gene products developed and maintained by the GO consortium. Other efforts in this field include the development of ontologies to more accurately describe expression data, for example, the work of the eVOC group *(119)*, and the developing Sequence Ontology *(120)* aimed at describing biological sequences.

Proteomics data are being produced on a large scale, and being lost on a large scale. Long lists of identifications are either published directly with the article, resulting in a voluminous and rather tedious read, or are included on the publisher's website as supplementary information. Genomics at today's scale would have to be conducted very differently if genomics data were not centrally archived and the requirement for data deposition prior to publication stipulated by scientific publishers and funding agencies. A number of proteomics databases exist among them, PRIDE *(121)*, GPM *(122)*, OPD *(123)*, PEDRo *(124)* and PeptideAtlas *(125)*. All of these repositories capture proteomics data and make it publicly available. Each of them, however, captures a different subset of data and provides it currently in a different form. Thus, accessing all proteomics data available in public databases still requires an enormous effort. To overcome this fragmentation of proteomics data and to establish a network of stable, synchronized proteomics resources, the ProteomExchange group was formed with the aim to establish a regular data exchange between major proteomics repositories.

Standardized, comprehensive deposition of proteomics data in public repositories will facilitate scientific collaboration and in large-scale collaborative projects, for example, in joint analysis of the same samples. This is particularly valuable in proteomics, where no single laboratory currently can elucidate the full proteome of a given system. The combination of different specialist proteomics technologies is essential to fully analyse a given system. Only the sharing of many different data sets created by diverse proteomics technologies will allow research on the system level.

While the global advantages of proteomics data standardization and deposition in public repositories are quite obvious, individual researchers might not consider them to be of immediate benefit. Data submission will engender a certain amount of extra effort, and easy public availability of the data might not always be seen as an advantage by the individual researcher. To minimize the effort for standardized data submission, it is necessary to engage commercial and open-source proteomics tool providers who will implement the standard data outputs in their products, thus minimizing the necessary effort for data conversion and submission. The participation of scientific publishers in such a process is also mandatory to ensure that the developed standards, tools and repositories meet the requirements of the publication process and will in turn promote the application of these standards by their authors, thus increasing the transparency, quality and public value of proteomics publications.

4. CONNECTING THE DATA: THE UNIPROT KNOWLEDGEBASE AND ITS HUMAN PROTEOMICS INITIATIVE, AND INTEGR8

Protein sequence information is generated in a highly redundant fashion resulting in sequence data corresponding to a single protein coming from many different sources such as genome projects, full-length cDNAs or less frequently protein sequencing. UniProt *(126)*, a resource unifying the Swiss-Prot *(127)*, TrEMBL *(127)* and PIR *(128)* databases, was created with the aim to provide a central database where sequences relating to the same protein are merged into a unique entry which describes all unique protein products of an individual gene. The UniProtKB provides extensive curated information on proteins, including functional annotation, classification and cross-references to external databases.

Sequence annotation provided in UniProtKB/Swiss-Prot, the manually curated branch of UniProtKB, comprises splice variants, polymorphisms, potential and experimentally confirmed sites of PTM, active sites, domains, cofactors and structure. Furthermore, information on biological pathways, catalytic activity, cofactors, regulation mechanisms, subcellular localization, tissue and developmental stage-specific expression of the protein and so on is provided. Extensive and increasing use of controlled vocabularies and ontologies such as the GO *(118)* and eVOC *(119)* improves data integration and programmatic data access.

UniProtKB/Swiss-Prot is supplemented by UniProtKB/TrEMBL containing protein sequences resulting from translation of nucleotide sequences in the EMBL sequence database, which are electronically annotated based on similarity to sequences in UniProtKB/Swiss-Prot. The UniProtKB contains a set of approximately 74,000 human sequences; however, this will include many splice variants, which will eventually be merged into a single entry within UniProtKB/Swiss-Prot. Currently, an average of about five nucleotide entries are used

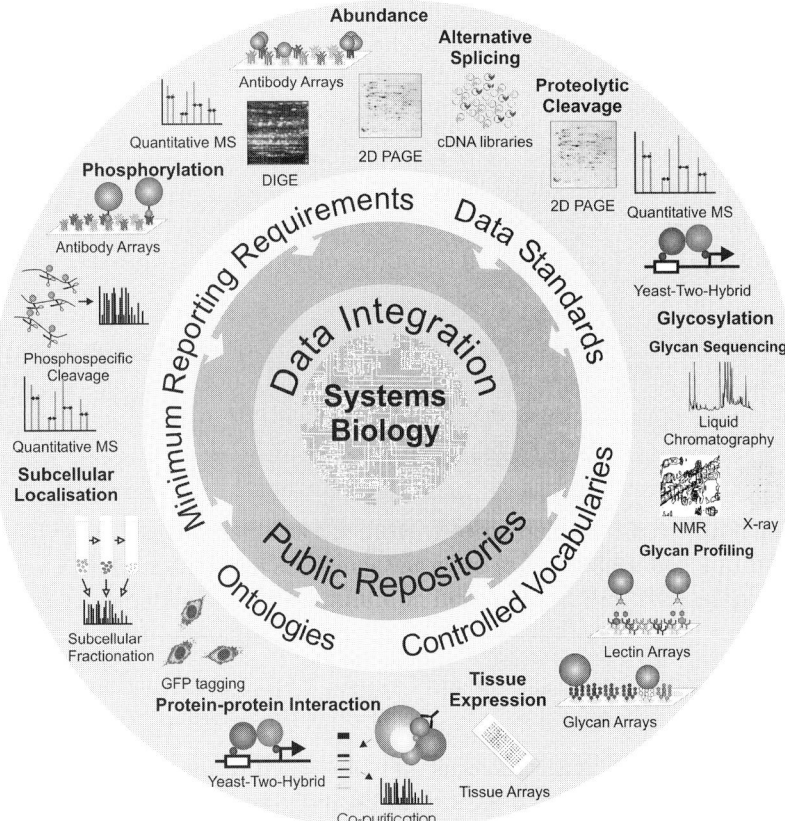

Fig. 2. Data integration. The Integr8 web portal offers easy access to integrated information on protein, transcript and genome level. The screen shot shows information on different isoforms of the tumour antigen p53 as provided by Integr8 including functional annotation based on Gene Ontology and InterPro cross-references, links to primary sequence databases, as well as various proteomics data repositories such as PRIDE, SWISS-2D-PAGE and IntAct.

to create one human UniProtKB/Swiss-Prot entry. These sequences can be further confirmed by protein sequencing either implicitly by the classical Edman sequencing technique or explicitly by mass spectrometry methods. Currently, more than 16% of the human entries contain such data.

UniProt has launched the Human Proteome Initiative (HPI) *(129)* a major project to annotate all known human sequences according to the Swiss-Prot quality standards. To date, 14,328 human protein records have been fully manually annotated with an additional 7887 splice variants being identified within these entries.

Strength of the UniProtKB is the extensive cross-referencing made to other, more specialized databases, offering a protein-centric way to move from an UniProtKB record to these different databases. Another approach to integrate many of these resources is to allow navigation from genome to gene to transcript to protein or the other way around. The Integr8 project *(130)* has been developed, as a joint effort between different European data providers, to attempt this. The Integr8 web portal provides easy access to integrated information about deciphered genomes and their corresponding transcripts and proteins. Available data include DNA sequences (from databases including the EMBL nucleotide sequence database, Genome Reviews and Ensembl); protein sequences (from databases including the UniProtKB and IPI); statistical genome and proteome analysis (performed using InterPro, CluSTr and GOA) and information about orthology, paralogy and synteny. Additional structural data are also available based on information derived from the Protein Data Bank and HSSP *(131)*. Figure 2 shows a screen shot of information on p53 as provided by the Integr8 website. Users can also configure their own analyses using interactive interfaces. The underlying data can be accessed for download via FTP and BioMart, a development of the EnsMart data warehousing system *(132)*.

5. CONCLUSION

We are still a long way from a full understanding of the human proteome, in particular of the specific role each molecule plays in the cellular context, but our knowledge is growing daily with increasing speed due to large-scale proteomics approaches and coordination of efforts through consortia and initiatives following the path shown by the Human Genome Project. A wealth of data are being generated on diverse aspects of protein function and made publicly accessible through an array of interlinked databases (See Table 1). Existing

Table 1
Proteomics Databases

Functional aspect	Resource	URL	Information
Alternative splicing	H-InvDB	http://www.h-invitational.jp	Full-length cDNAs
	ASD	http://www.ebi.ac.uk/asd	Alternative splicing events
	Ensembl	http://www.ensembl.org	Splice variants
Phosphorylation	PhosphoSite	http://www.phosphosite.org	Phosphorylation sites
	Phoshpo.ELM	http://phospho.elm.eu.org	Phosphorylation sites
	Protein Kinase Resource	http://www.kinasenet.org	Protein kinases
	Kinase Pathway Database	http://kinasedb.ontology.ims.u-tokyo.ac.jp	Protein kinases
Proteolytic cleavage	MEROPS	http://merops.sanger.ac.uk	Proteases and protease inhibitors
Glycosylation	KEGG Glycan database	http://www.genome.jp/kegg/glycan	Glycan structures
	glycosciences.de	http://www.glycosciences.de	Glycan structures
Subcellular localization	LIFEdb	http://www.dkfz.de/LIFEdb	Protein localization, interaction, functional assays and expression
	GFP-cDNA localisation project	http://gfp-cdna.embl.de	Subcellular localization of GFP-tagged proteins

	Organelle DB	http://organelledb.lsi.umich.edu	Mitochondrail proteins
	MitoRes	http://www2.ba.itb.cnr.it/MitoRes	Mitochondrail proteins
	Mitomap	http://www.mitomap.org	Nuclear proteins
	Nuclear Protein Database	http://npd.hgu.mrc.ac.uk/	
	NMP-db	http://www.rostlab.org/db/nmpdb/	Nuclear matrix proteins
	Hera database	http://www.mcb.mcgill.ca/~hera	ER proteins
Protein–protein interaction	BIND	http://www.blueprint.org	Protein interaction
	DIP	http://dip.doe-mbi.ucla.edu	Protein interaction
	IntAct	http://www.ebi.ac.uk/intact	Protein interaction
	MINT	http://mint.bio.uniroma2.it/mint	Protein interaction
	MIPS	http://mips.gsf.de/proj/ppi/	Protein interaction
	HPID	http://wilab.inha.ac.kr/hpid/	Protein interaction
	InterDom	http://interdom.lit.org.sg	Protein interaction
Tissue expression	Human Protein Atlase	http://www.proteinatlas.org	Tissue expression
Integrating resources	UniProtKB	http://www.uniprot.org	Curated information on proteins
	Integr8	http://www.ebi.ac.uk/integr8	Integrated information about deciphered genomes and their corresponding proteomes

and emerging data standards will facilitate access and exchange of these disperse data. This is a major prerequisite for the next stage in post-genome research, the integration of data from different 'omics' domains to enable studies
on the systems level which will eventually facilitate the discovery of new targets for cancer therapy, and provide a better understanding of the implications therapeutic intervention will have on the system as a whole.

REFERENCES

1. Dulbecco R. A turning point in cancer research: sequencing the human genome. Science 1986; 231:1055–6.
2. Stein L. Genome annotation: from sequence to biology. Nat Rev Genet 2001; 2:493–503.
3. Lander ES, Linton LM, Birren B, et al. Initial sequencing and analysis of the human genome. Nature 2001; 409:860–921.
4. Venter JC, Adams MD, Myers EW, et al. The sequence of the human genome. Science 2001; 291:1304–51.
5. IHGSC. Finishing the euchromatic sequence of the human genome. Nature 2004; 431:931–45.
6. Byrd MP, Zamora M, Lloyd RE. Generation of multiple isoforms of eukaryotic translation initiation factor 4GI by use of alternate translation initiation codons. Mol Cell Biol 2002; 22:4499–511.
7. Humphery-Smith I. A human proteome project with a beginning and an end. Proteomics 2004; 4:2519–21.
8. Stanford DR, Martin NC, Hopper AK. ADEPTs: information necessary for subcellular distribution of eukaryotic sorting isozymes resides in domains missing from eubacterial and archaeal counterparts. Nucleic Acids Res 2000; 28:383–92.
9. Resch A, Xing Y, Modrek B, Gorlick M, Riley R, Lee C. Assessing the impact of alternative splicing on domain interactions in the human proteome. J Proteome Res 2004; 3:76–83.
10. Nakao M, Barrero RA, Mukai Y, Motono C, Suwa M, Nakai K. Large-scale analysis of human alternative protein isoforms: pattern classification and correlation with subcellular localization signals. Nucleic Acids Res 2005; 33:2355–63.
11. Modrek B, Resch A, Grasso C, Lee C. Genome-wide detection of alternative splicing in expressed sequences of human genes. Nucleic Acids Res 2001; 29:2850–9.
12. Johnson JM, Castle J, Garrett-Engele P, et al. Genome-wide survey of human alternative pre-mRNA splicing with exon junction microarrays. Science 2003; 302:2141–4.
13. Krawczak M, Reiss J, Cooper DN. The mutational spectrum of single base-pair substitutions in mRNA splice junctions of human genes: causes and consequences. Hum Genet 1992; 90:41–54.
14. Imanishi T, Itoh T, Suzuki Y, et al. Integrative annotation of 21,037 human genes validated by full-length cDNA clones. PLoS Biol 2004; 2:e162.
15. Thanaraj TA, Stamm S, Clark F, Riethoven JJ, Le Texier V, Muilu J. ASD: the Alternative Splicing Database. Nucleic Acids Res 2004; 32:D64–9.

16. Zhou H, Watts JD, Aebersold R. A systematic approach to the analysis of protein phosphorylation. Nat Biotechnol 2001; 19:375–8.
17. Ficarro SB, McCleland ML, Stukenberg PT, et al. Phosphoproteome analysis by mass spectrometry and its application to Saccharomyces cerevisiae. Nat Biotechnol 2002; 20:301–5.
18. Knight ZA, Schilling B, Row RH, Kenski DM, Gibson BW, Shokat KM. Phosphospecific proteolysis for mapping sites of protein phosphorylation. Nat Biotechnol 2003; 21:1047–54.
19. Grubb RL, Calvert VS, Wulkuhle JD, et al. Signal pathway profiling of prostate cancer using reverse phase protein arrays. Proteomics 2003; 3:2142–6.
20. Gembitsky DS, Lawlor K, Jacovina A, Yaneva M, Tempst P. A prototype antibody microarray platform to monitor changes in protein tyrosine phosphorylation. Mol Cell Proteomics 2004; 3:1102–18.
21. Tao WA, Wollscheid B, O'Brien R, et al. Quantitative phosphoproteome analysis using a dendrimer conjugation chemistry and tandem mass spectrometry. Nat Methods 2005; 2:591–8.
22. Cantin GT, Venable JD, Cociorva D, Yates JR, III. Quantitative phosphoproteomic analysis of the tumor necrosis factor pathway. J Proteome Res 2006; 5:127–34.
23. Blagoev B, Ong SE, Kratchmarova I, Mann M. Temporal analysis of phosphotyrosine-dependent signaling networks by quantitative proteomics. Nat Biotechnol 2004; 22:1139–45.
24. Ballif BA, Roux PP, Gerber SA, MacKeigan JP, Blenis J, Gygi SP. Quantitative phosphorylation profiling of the ERK/p90 ribosomal S6 kinase-signaling cassette and its targets, the tuberous sclerosis tumor suppressors. Proc Natl Acad Sci USA 2005; 102:667–72.
25. Hornbeck PV, Chabra I, Kornhauser JM, Skrzypek E, Zhang B. PhosphoSite: a bioinformatics resource dedicated to physiological protein phosphorylation. Proteomics 2004; 4:1551–61.
26. Diella F, Cameron S, Gemund C, et al. Phospho.ELM: a database of experimentally verified phosphorylation sites in eukaryotic proteins. BMC Bioinformatics 2004; 5:79.
27. Niedner RH, Buzko OV, Haste NM, Taylor A, Gribskov M, Taylor SS. Protein kinase resource: an integrated environment for phosphorylation research. Proteins 2006; 63:78–86.
28. Koike A, Kobayashi Y, Takagi T. Kinase pathway database: an integrated protein-kinase and NLP-based protein-interaction resource. Genome Res 2003; 13: 1231–43.
29. Manning G, Whyte DB, Martinez R, Hunter T, Sudarsanam S. The protein kinase complement of the human genome. Science 2002; 298:1912–34.
30. Apweiler R, Hermjakob H, Sharon N. On the frequency of protein glycosylation, as deduced from analysis of the SWISS-PROT database. Biochim Biophys Acta 1999; 1473:4–8.
31. Liu D, Shriver Z, Venkataraman G, El Shabrawi Y, Sasisekharan R. Tumor cell surface heparan sulfate as cryptic promoters or inhibitors of tumor growth and metastasis. Proc Natl Acad Sci USA 2002; 99:568–73.
32. Fuster MM, Brown JR, Wang L, Esko JD. A disaccharide precursor of sialyl Lewis X inhibits metastatic potential of tumor cells. Cancer Res 2003; 63: 2775–81.

33. Ishida H, Togayachi A, Sakai T, et al. A novel beta1,3-N-acetylglucosaminyl transferase (beta3Gn-T8), which synthesizes poly-N-acetyllactosamine, is dramatically upregulated in colon cancer. FEBS Lett 2005; 579: 71–8.
34. Dube DH, Bertozzi CR. Glycans in cancer and inflammation–potential for therapeutics and diagnostics. Nat Rev Drug Discov 2005; 4:477–88.
35. Dell A, Morris HR. Glycoprotein structure determination by mass spectrometry. Science 2001; 291:2351–6.
36. Joshi HJ, Harrison MJ, Schulz BL, Cooper CA, Packer NH, Karlsson NG. Development of a mass fingerprinting tool for automated interpretation of oligosaccharide fragmentation data. Proteomics 2004; 4:1650–64.
37. Morelle W, Flahaut C, Michalski JC, Louvet A, Mathurin P, Klein A. Mass spectrometric approach for screening modifications of total serum N-glycome in human diseases: application to cirrhosis. Glycobiology 2006; 16:281–93.
38. Ethier M, Saba JA, Spearman M, et al. Application of the StrOligo algorithm for the automated structure assignment of complex N-linked glycans from glycoproteins using tandem mass spectrometry. Rapid Commun Mass Spectrom 2003; 17:2713–20.
39. Manzi AE, Norgard-Sumnicht K, Argade S, Marth JD, van Halbeek H, Varki A. Exploring the glycan repertoire of genetically modified mice by isolation and profiling of the major glycan classes and nano-NMR analysis of glycan mixtures. Glycobiology 2000; 10:669–89.
40. Rudd PM, Colominas C, Royle L, et al. A high-performance liquid chromatography based strategy for rapid, sensitive sequencing of N-linked oligosaccharide modifications to proteins in sodium dodecyl sulphate polyacrylamide electrophoresis gel bands. Proteomics 2001; 1:285–94.
41. Hashii N, Kawasaki N, Itoh S, Hyuga M, Kawanishi T, Hayakawa T. Glycomic/glycoproteomic analysis by liquid chromatography/mass spectrometry: analysis of glycan structural alteration in cells. Proteomics 2005; 5:4665–72.
42. Blixt O, Head S, Mondala T, et al. Printed covalent glycan array for ligand profiling of diverse glycan binding proteins. Proc Natl Acad Sci USA 2004; 101:17033–8.
43. Xia B, Kawar ZS, Ju T, Alvarez RA, Sachdev GP, Cummings RD. Versatile fluorescent derivatization of glycans for glycomic analysis. Nat Methods 2005; 2:845–50.
44. Pilobello KT, Krishnamoorthy L, Slawek D, Mahal LK. Development of a lectin microarray for the rapid analysis of protein glycopatterns. Chembiochem 2005; 6:985–9.
45. Zheng T, Peelen D, Smith LM. Lectin arrays for profiling cell surface carbohydrate expression. J Am Chem Soc 2005; 127:9982–3.
46. Kuno A, Uchiyama N, Koseki-Kuno S, et al. Evanescent-field fluorescence-assisted lectin microarray: a new strategy for glycan profiling. Nat Methods 2005; 2:851–6.
47. Hashimoto K, Goto S, Kawano S, et al. KEGG as a glycome informatics resource. Glycobiology 2005.
48. Sahoo SS, Thomas C, Sheth A, Henson C, York WS. GLYDE-an expressive XML standard for the representation of glycan structure. Carbohydr Res 2005; 340:2802–7.
49. Bohne-Lang A, Lang E, Forster T, von der Lieth CW. LINUCS: linear notation for unique description of carbohydrate sequences. Carbohydr Res 2001; 336: 1–11.

50. Kikuchi N, Kameyama A, Nakaya S, et al. The carbohydrate sequence markup language (CabosML): an XML description of carbohydrate structures. Bioinformatics 2005; 21:1717–8.
51. Neurath H. Proteolytic enzymes, past and future. Proc Natl Acad Sci USA 1999; 96:10962–3.
52. Hotary KB, Allen ED, Brooks PC, Datta NS, Long MW, Weiss SJ. Membrane type I matrix metalloproteinase usurps tumor growth control imposed by the three-dimensional extracellular matrix. Cell 2003; 114:33–45.
53. McCawley LJ, Matrisian LM. Matrix metalloproteinases: multifunctional contributors to tumor progression. Mol Med Today 2000; 6:149–56.
54. Puente XS, Sanchez LM, Overall CM, Lopez-Otin C. Human and mouse proteases: a comparative genomic approach. Nat Rev Genet 2003; 4:544–58.
55. Rawlings ND, Morton FR, Barrett AJ. MEROPS: the peptidase database. Nucleic Acids Res 2006; 34:D270–2.
56. Hwang IK, Park SM, Kim SY, Lee ST. A proteomic approach to identify substrates of matrix metalloproteinase-14 in human plasma. Biochim Biophys Acta 2004; 1702:79–87.
57. Overall CM, Tam EM, Kappelhoff R, et al. Protease degradomics: mass spectrometry discovery of protease substrates and the CLIP-CHIP, a dedicated DNA microarray of all human proteases and inhibitors. Biol Chem 2004; 385:493–504.
58. Lee AY, Park BC, Jang M, et al. Identification of caspase-3 degradome by two-dimensional gel electrophoresis and matrix-assisted laser desorption/ionization-time of flight analysis. Proteomics 2004; 4:3429–36.
59. Dohi T, Beltrami E, Wall NR, Plescia J, Altieri DC. Mitochondrial survivin inhibits apoptosis and promotes tumorigenesis. J Clin Invest 2004; 114:1117–27.
60. Simpson JC, Wellenreuther R, Poustka A, Pepperkok R, Wiemann S. Systematic subcellular localization of novel proteins identified by large-scale cDNA sequencing. EMBO Rep 2000; 1:287–92.
61. Liebel U, Starkuviene V, Erfle H, et al. A microscope-based screening platform for large-scale functional protein analysis in intact cells. FEBS Lett 2003; 554:394–8.
62. Bannasch D, Mehrle A, Glatting KH, Pepperkok R, Poustka A, Wiemann S. LIFEdb: a database for functional genomics experiments integrating information from external sources, and serving as a sample tracking system. Nucleic Acids Res 2004; 32:D505–8.
63. Andersen JS, Lyon CE, Fox AH, et al. Directed proteomic analysis of the human nucleolus. Curr Biol 2002; 12:1–11.
64. Andersen JS, Lam YW, Leung AK, et al. Nucleolar proteome dynamics. Nature 2005; 433:77–83.
65. Schirmer EC, Florens L, Guan T, Yates JR, III, Gerace L. Nuclear membrane proteins with potential disease links found by subtractive proteomics. Science 2003; 301:1380–2.
66. Breuza L, Halbeisen R, Jeno P, et al. Proteomics of endoplasmic reticulum-Golgi intermediate compartment (ERGIC) membranes from brefeldin A-treated HepG2 cells identifies ERGIC-32, a new cycling protein that interacts with human Erv46. J Biol Chem 2004; 279:47242–53.
67. Bell AW, Ward MA, Blackstock WP, et al. Proteomics characterization of abundant Golgi membrane proteins. J Biol Chem 2001; 276:5152–65.

68. Pisitkun T, Shen RF, Knepper MA. Identification and proteomic profiling of exosomes in human urine. Proc Natl Acad Sci USA 2004; 101:13368–73.
69. Mears R, Craven RA, Hanrahan S, et al. Proteomic analysis of melanoma-derived exosomes by two-dimensional polyacrylamide gel electrophoresis and mass spectrometry. Proteomics 2004; 4:4019–31.
70. Andersen JS, Wilkinson CJ, Mayor T, Mortensen P, Nigg EA, Mann M. Proteomic characterization of the human centrosome by protein correlation profiling. Nature 2003; 426:570–4.
71. Sauer G, Korner R, Hanisch A, Ries A, Nigg EA, Sillje HH. Proteome analysis of the human mitotic spindle. Mol Cell Proteomics 2005; 4:35–43.
72. Taylor SW, Fahy E, Zhang B, et al. Characterization of the human heart mitochondrial proteome. Nat Biotechnol 2003; 21:281–6.
73. Wiwatwattana N, Kumar A. Organelle DB: a cross-species database of protein localization and function. Nucleic Acids Res 2005; 33:D598–604.
74. Catalano D, Licciulli F, Turi A, Grillo G, Saccone C, D'Elia D. MitoRes: a resource of nuclear-encoded mitochondrial genes and their products in Metazoa. BMC Bioinformatics 2006; 7:36.
75. Brandon MC, Lott MT, Nguyen KC, et al. MITOMAP: a human mitochondrial genome database–2004 update. Nucleic Acids Res 2005; 33:D611–3.
76. Dellaire G, Farrall R, Bickmore WA. The Nuclear Protein Database (NPD): sub-nuclear localisation and functional annotation of the nuclear proteome. Nucleic Acids Res 2003; 31:328–30.
77. Mika S, Rost B. NMPdb: Database of nuclear matrix proteins. Nucleic Acids Res 2005; 33:D160–3.
78. Scott M, Lu G, Hallett M, Thomas DY. The Hera database and its use in the characterization of endoplasmic reticulum proteins. Bioinformatics 2004; 20:937–44.
79. Su AI, Wiltshire T, Batalov S, et al. A gene atlas of the mouse and human protein-encoding transcriptomes. Proc Natl Acad Sci USA 2004; 101:6062–7.
80. Hanash S. HUPO initiatives relevant to clinical proteomics. Mol Cell Proteomics 2004; 3:298–301.
81. Uhlen M, Ponten F. Antibody-based proteomics for human tissue profiling. Mol Cell Proteomics 2005; 4:384–93.
82. Uhlen M, Bjorling E, Agaton C, et al. A human protein atlas for normal and cancer tissues based on antibody proteomics. Mol Cell Proteomics 2005; 4:1920–32.
83. Warford A, Howat W, McCafferty J. Expression profiling by high-throughput immunohistochemistry. J Immunol Methods 2004; 290:81–92.
84. Warford A. Tissue microarrays: fast-tracking protein expression at the cellular level. Expert Rev Proteomics 2004; 1:283–92.
85. Lindskog M, Rockberg J, Uhlen M, Sterky F. Selection of protein epitopes for antibody production. Biotechniques 2005; 38:723–7.
86. Nilsson P, Paavilainen L, Larsson K, et al. Towards a human proteome atlas: high-throughput generation of mono-specific antibodies for tissue profiling. Proteomics 2005; 5:4327–37.
87. Gygi SP, Rochon Y, Franza BR, Aebersold R. Correlation between protein and mRNA abundance in yeast. Mol Cell Biol 1999; 19:1720–30.
88. Tonge R, Shaw J, Middleton B, et al. Validation and development of fluorescence two-dimensional differential gel electrophoresis proteomics technology. Proteomics 2001; 1:377–96.

89. Gygi SP, Rist B, Gerber SA, Turecek F, Gelb MH, Aebersold R. Quantitative analysis of complex protein mixtures using isotope-coded affinity tags. Nat Biotechnol 1999; 17:994–9.
90. Ong SE, Mann M. Mass spectrometry-based proteomics turns quantitative. Nat Chem Biol 2005; 1:252–62.
91. Beynon RJ, Doherty MK, Pratt JM, Gaskell SJ. Multiplexed absolute quantification in proteomics using artificial QCAT proteins of concatenated signature peptides. Nat Methods 2005; 2:587–9.
92. Haab BB, Dunham MJ, Brown PO. Protein microarrays for highly parallel detection and quantitation of specific proteins and antibodies in complex solutions. Genome Biol 2001; 2:RESEARCH0004.
93. Paweletz CP, Charboneau L, Bichsel VE, et al. Reverse phase protein microarrays which capture disease progression show activation of pro-survival pathways at the cancer invasion front. Oncogene 2001; 20:1981–9.
94. Nishizuka S, Charboneau L, Young L, et al. Proteomic profiling of the NCI-60 cancer cell lines using new high-density reverse-phase lysate microarrays. Proc Natl Acad Sci USA 2003; 100:14229–34.
95. Sreekumar A, Nyati MK, Varambally S, et al. Profiling of cancer cells using protein microarrays: discovery of novel radiation-regulated proteins. Cancer Res 2001; 61:7585–93.
96. Miller JC, Zhou H, Kwekel J, et al. Antibody microarray profiling of human prostate cancer sera: antibody screening and identification of potential biomarkers. Proteomics 2003; 3:56–63.
97. Nishizuka S. Profiling cancer stem cells using protein array technology. Eur J Cancer 2006;42(9):1273–82.
98. Uetz P, Giot L, Cagney G, et al. A comprehensive analysis of protein-protein interactions in Saccharomyces cerevisiae. Nature 2000; 403:623–7.
99. Ito T, Chiba T, Ozawa R, Yoshida M, Hattori M, Sakaki Y. A comprehensive two-hybrid analysis to explore the yeast protein interactome. Proc Natl Acad Sci USA 2001; 98:4569–74.
100. Giot L, Bader JS, Brouwer C, et al. A protein interaction map of Drosophila melanogaster. Science 2003; 302:1727–36.
101. Li S, Armstrong CM, Bertin N, et al. A map of the interactome network of the metazoan C. elegans. Science 2004; 303:540–3.
102. Rual JF, Venkatesan K, Hao T, et al. Towards a proteome-scale map of the human protein-protein interaction network. Nature 2005; 437:1173–8.
103. Stelzl U, Worm U, Lalowski M, et al. A human protein-protein interaction network: a resource for annotating the proteome. Cell 2005; 122:957–68.
104. Bouwmeester T, Bauch A, Ruffner H, et al. A physical and functional map of the human TNF-alpha/NF-kappa B signal transduction pathway. Nat Cell Biol 2004; 6:97–105.
105. Falsone SF, Gesslbauer B, Tirk F, Piccinini AM, Kungl AJ. A proteomic snapshot of the human heat shock protein 90 interactome. FEBS Lett 2005; 579:6350–4.
106. Blagoev B, Kratchmarova I, Ong SE, Nielsen M, Foster LJ, Mann M. A proteomics strategy to elucidate functional protein-protein interactions applied to EGF signaling. Nat Biotechnol 2003; 21:315–8.
107. Colland F, Jacq X, Trouplin V, et al. Functional proteomics mapping of a human signaling pathway. Genome Res 2004; 14:1324–32.
108. de Lichtenberg U, Jensen LJ, Brunak S, Bork P. Dynamic complex formation during the yeast cell cycle. Science 2005; 307:724–7.

109. Barrios-Rodiles M, Brown KR, Ozdamar B, et al. High-throughput mapping of a dynamic signaling network in mammalian cells. Science 2005; 307:1621–5.
110. Hermjakob H, Montecchi-Palazzi L, Lewington C, et al. IntAct: an open source molecular interaction database. Nucleic Acids Res 2004; 32:D452–5.
111. Bader GD, Betel D, Hogue CW. BIND: the Biomolecular Interaction Network Database. Nucleic Acids Res 2003; 31:248–50.
112. Zanzoni A, Montecchi-Palazzi L, Quondam M, Ausiello G, Helmer-Citterich M, Cesareni G. MINT: a Molecular INTeraction database. FEBS Lett 2002; 513: 135–40.
113. Salwinski L, Miller CS, Smith AJ, Pettit FK, Bowie JU, Eisenberg D. The Database of Interacting Proteins: 2004 update. Nucleic Acids Res 2004; 32: D449–51.
114. Stark C, Breitkreutz BJ, Reguly T, Boucher L, Breitkreutz A, Tyers M. BioGRID: a general repository for interaction datasets. Nucleic Acids Res 2006; 34:D535–9.
115. Carr S, Aebersold R, Baldwin M, Burlingame A, Clauser K, Nesvizhskii A. The need for guidelines in publication of peptide and protein identification data: Working Group on Publication Guidelines for Peptide and Protein Identification Data. Mol Cell Proteomics 2004; 3:531–3.
116. Taylor CF, Paton NW, Garwood KL, et al. A systematic approach to modeling, capturing, and disseminating proteomics experimental data. Nat Biotechnol 2003; 21:247–54.
117. Pedrioli PG, Eng JK, Hubley R, et al. A common open representation of mass spectrometry data and its application to proteomics research. Nat Biotechnol 2004; 22:1459–66.
118. Ashburner M, Ball CA, Blake JA, et al. Gene ontology: tool for the unification of biology. The Gene Ontology Consortium. Nat Genet 2000; 25:25–9.
119. Kelso J, Visagie J, Theiler G, et al. eVOC: a controlled vocabulary for unifying gene expression data. Genome Res 2003; 13:1222–30.
120. Eilbeck K, Lewis SE, Mungall CJ, et al. The Sequence Ontology: a tool for the unification of genome annotations. Genome Biol 2005; 6:R44.
121. Jones P, Cote RG, Martens L, et al. PRIDE: a public repository of protein and peptide identifications for the proteomics community. Nucleic Acids Res 2006; 34:D659–63.
122. Craig R, Cortens JP, Beavis RC. Open source system for analyzing, validating, and storing protein identification data. J Proteome Res 2004; 3:1234–42.
123. Prince JT, Carlson MW, Wang R, Lu P, Marcotte EM. The need for a public proteomics repository. Nat Biotechnol 2004; 22:471–2.
124. Garwood K, McLaughlin T, Garwood C, et al. PEDRo: a database for storing, searching and disseminating experimental proteomics data. BMC Genomics 2004; 5:68.
125. Desiere F, Deutsch EW, King NL, et al. The PeptideAtlas project. Nucleic Acids Res 2006; 34:D655–8.
126. Apweiler R, Bairoch A, Wu CH, et al. UniProt: the universal protein knowledgebase. Nucleic Acids Res 2004; 32:D115–9.
127. Boeckmann B, Bairoch A, Apweiler R, et al. The SWISS-PROT protein knowledgebase and its supplement TrEMBL in 2003. Nucleic Acids Res 2003; 31: 365–70.
128. Wu CH, Yeh LS, Huang H, et al. The protein information resource. Nucleic Acids Res 2003; 31:345–7.

129. O'Donovan C, Apweiler R, Bairoch A. The human proteomics initiative (HPI). Trends Biotechnol 2001; 19:178–81.
130. Kersey P, Bower L, Morris L, et al. Integr8 and Genome reviews: integrated views of complete genomes and proteomes. Nucleic Acids Res 2005; 33: D297–302.
131. Dodge C, Schneider R, Sander C. The HSSP database of protein structure-sequence alignments and family profiles. Nucleic Acids Res 1998; 26:313–5.
132. Kasprzyk A, Keefe D, Smedley D, et al. EnsMart: a generic system for fast and flexible access to biological data. Genome Res 2004; 14:160–9.

11 Regulatory Issues in the Co-Development of Oncology Drugs and Proteomic Tests: An Overview

Dave Li, Joseph Hackett, Maria Chan, Gene Pennello, and Steve Gutman

CONTENTS

1. INTRODUCTION
2. THE REGULATORY PATHWAYS TO MARKET
3. THE CLASSIFICATION OF TUMOR MARKERS
4. FDA OUTREACH ACTIVITIES
5. SOME CRITICAL ISSUES WITH REGULATORY IMPLICATIONS FOR PROTEOMIC CANCER BIOMARKERS
6. CONCLUSION

SUMMARY

In this article, we review some relevant issues in cancer proteomic biomarker development and validation for targeted cancer therapeutics with emphasis on the importance of developing a clearly stated intended use and indication for use, establishing a

From: *Cancer Drug Discovery and Development*
Cancer Proteomics: From Bench to Bedside
Edited by: S. S. Daoud © Humana Press Inc., Totowa, NJ

reliable laboratory measurement, using robust clinical study designs with representative patient populations and effective controls, and addressing issues of biological validations and potential bias associated with predictive biomarker validation.

Key Words: Cancer; proteomics; target; therapy; validation; regulation

1. INTRODUCTION

The completion of the human genome project in parallel with the advent of new developments in diagnostic technology has clearly revolutionized modern laboratory medicine. It is now quite feasible to utilize patient-specific genomic or proteomic information to allow for refined treatment decision making. In particular, introduction of the protein chip, refined methods for use of mass spectroscopy, information systems to allow for pattern recognition, large-scale protein identification, novel data processing, and bioinformatics analysis systems have brought the search for proteomic markers to the forefront.

The introduction of both genomic and proteomic markers has been accompanied by a marked shift in the traditional paradigm applied to the search for new biomarkers. Traditionally, biomarker discovery has been hypothesis driven. Research activities were largely directed to one particular issue at a time, focused on testing a specific hypothesis derived using clinical observation, limited experimental data, or informed modeling. With the delineation of the human genomic map, multiple potential drug targets can be identified at one time, and the search for new drug targets has evolved increasingly toward use of empiric database-driven markers *(1)*. This enables the generation of large numbers of potential targets without specific working hypotheses or concrete mechanistic insights in a relatively short time *(2,3)*. The new approach is providing opportunities for new biomarker discovery, although interpretation of the data generated remains a challenge to biomedical researchers.

In the context of oncology drug development, biomarkers can be categorized into those used to confirm pharmacological mechanisms of action, to demonstrate biological mechanisms of action, and/or to predict clinical outcome. Use of protein biomarkers in the laboratory can produce unique challenges. Unlike nucleic acids, proteins are difficult to manipulate in a laboratory setting due to the fact that they cannot be fully duplicated even with a DNA template. Mass spectrometry coupled with separation technology and informatics has dramatically improved our capability to study and use proteins in new biomarker and target discovery *(4,5)*. Recent work with targeted therapy has shown that measurement of the genetic variations of a molecular target can predict treatment responses in specific subsets of patient populations *(6)*. In addition, developments in the area of functional proteomics are expected to aid in understanding the mechanisms of drug resistance. Protein biomarkers may be particularly valuable in guiding targeted therapeutics because in most cases they may identify direct targets of the drug actions with the requisite post-translational modifications required for appropriate functions.

It becomes evident that the high dimensionality of proteomics data poses a challenge in statistical analysis to assure the biological verification and validation of a new biomarker *(7,8)*. For instance, particular care is needed to control the false discovery rate inherent in use of the supervised machine learning methods when applied to a large pool of candidate biomarkers with limited patient samples. Careful and creative design is needed to assure that the fundamental performance of a new marker is robust and reproducible rather than artifactual. In addition, the study design needs to further demonstrate that the biomarker provides added values to existing diagnostic modalities and therefore has clinical utility. High dimensionality in data presents unique study challenges, especially when the dimensionality of the biomarker pool greatly exceeds the number of samples studied. But this high dimensionality is likely to be of unique value in allowing a diagnostic to move from the limited information provided by unique markers to the use of a more nuanced and sophisticated approach based on the systemic biology of the diseases being studied. Given the biological complexity of cancers, this approach seems particularly appropriate.

From the FDA's perspective, new proteomic technology, like any new technology, is based on the generation of good scientific information collected using good laboratory practices and applied using good manufacturing systems consistent with FDA's quality system regulations (Title 21 Code of Federal Regulations FDA, Subchapter H, Part 820 Quality Systems Regulations). Ideally, the analytes should be specific, with unique tissue or organ distribution, be clinically relevant to patient outcome, and the measurement be standardized and reproducible. Analytical and clinical data should be transparent and scientifically and clinically verifiable *(9–12)*.

Proteomics is an evolving science. While FDA has experience in regulating cancer biomarkers, its experience in regulating the proteomic cancer biomarkers and their applications in targeted cancer therapy are limited. It is our hope that by working together with cancer researchers, industry, and oncologists, we will be able to facilitate the rapid transfer of new technology from the research bench to the patient bedside. In this chapter, we will provide a brief description of current framework and regulatory pathways for in vitro diagnostic devices (IVD) devices, more commonly referred to as laboratory tests, and point out some critical issues with regulatory implications pertinent to diagnostic proteomics in predicting the safety and effectiveness of cancer therapeutics.

2. THE REGULATORY PATHWAYS TO MARKET

FDA has regulatory oversight over all Medical Devices, which include IVDs. FDA regulates tests or test systems for use in clinical laboratories, physician offices, or other point of care settings.

Tests regulated by FDA are categorized into one of three classes (class I, II, and III). FDA exerts its lowest level of oversight to class I devices, medium oversight to class II devices, and its highest level of oversight to class III devices. Manufacturers desiring to commercially market a laboratory test seek pre-market approval (PMA) for class III or pre-market notification (510K) for class II from FDA.

The most commonly used route is the pre-market notification, described in section 510, subpart k of the Federal Food, Drug, and Cosmetic Act. This submission is commonly referred to as a 510(k). The new test kit or system can be provided in a variety of configurations and may include hardware, software, reagents, and/or other accessories along with instructions for use and clear labeling as appropriate. Most commonly, new 510(k) submissions are reviewed by comparing to an existing device which was on the market at the time of the 1976 Medical Device Amendments or has been cleared since 1976 as being equivalent to one of these on the market in 1976 (predicate device).

The new device must be shown to be substantially equivalent to the identified predicate device. FDA is allotted a 90-day review time and attempts to work interactively with the sponsor or manufacturer to meet this goal.

A second route for FDA to bring a new product to market is the PMA Application (PMA). PMAs are associated with tests for which there is a high risk to the patient from the laboratory results, or where little is known about the test. These are considered class III devices, and FDA uses the term pre-market approval to describe its process for these applications. These products are not studied in comparison to a predicate device, but instead are shown to be safe and effective using appropriate clinical and/or laboratory data.

Safety is focused on assuring the benefits of use outweigh the risks. Effectiveness is focused on assuring a new device provides clinically significant results. The PMA review process allows for a 180-day review timeline. For novel or complex technology, FDA often holds formal advisory committee meetings to assure outside scientific expertise is applied to its review process and that a public and transparent discussion and review of the technology is effected.

In spite of differing administrative and scientific regulatory thresholds applied to these two types of submissions, there is varying overlap in the core information that may be required to complete premarket review. Performance data for both submission types should demonstrate the accuracy of test performance when compared to a predicate device or appropriate clinical and/or laboratory data. In addition, test precision, analytical specificity, and, when appropriate, other aspects of detection or measurement are also reviewed.

While there is no single path for demonstrating clinical performance of an assay, roadmaps for this type of analysis have been published *(10,13,14)*. In addition, numerous voluntary standards (http://www.clsi.org) and FDA guidances (http://www.fda.gov/cdrh/oivd/regulatory-guidance.html) provide insight into this activity. Sponsors can provide information either based on their own clinical studies of the intended use population or in some cases may cite relevant published clinical studies or utilize data derived from other external sources.

In addition as part of the PMA review process, the manufacturing site for the new device must pass an inspection by FDA. The inspection is to assure that the manufacturer can consistently produce the product to meet its own written specifications. The FDA inspects the test manufacturer's facility and manufacturing specifications, as well as records generated by the manufacturer in the production of lots or batches of the test or test system. Again, the intent is to ensure that a manufacturing system is in place to provide for consistent production of the product over time in conformance with product claims and labeling.

FDA pre-market work is performed in a highly transparent manner. FDA device reviews for cleared IVD 510(k) submissions are posted on the Office of In Vitro Diagnostic Device (OIVD) web page (http://www.fda.gov/cdrh/oivd) using a standardized review template. A summary of safety and effectiveness for approved PMA submissions is posted as well.

3. THE CLASSIFICATION OF TUMOR MARKERS

FDA began regulating tumor-associated antigen test systems in 1973. Because false-positive and false-negative results could result in major errors in treatment, tumor-associated antigens were originally considered high-risk products and designated class III devices subject to review as PMA submissions. Although the menu of tests approved for market was small (carcinoembryonic antigen, alpha fetoprotein, and prostate-specific antigen), more then two dozen applications were approved for market over the next 25 years. In 1999, FDA received a petition to down-classify IVDs used to monitor cancer patients from class III to class II devices. This petition was based on the argument that monitoring tests provided a lower risk than screening tests because the patient had already received a cancer diagnosis. It was also based on the belief that after 25 years of common use, these monitoring tests and the possibility of false-positive and false-negative results were well understood by health care professionals treating cancer. As a result, IVDs for use in testing for cancer were divided into high risk class III screening and diagnostic devices and moderate risk class II monitoring devices.

4. FDA OUTREACH ACTIVITIES

In an effort to better understand new technologies like genomics and proteomics, the agency has established several outreach activities. The term "informal" is used to designate submissions that are voluntarily submitted to the agency, primarily for educational purposes. These submissions are not used to obtain regulatory decisions, but serve as a means of communication between the agency and industry.

The Center for Drug Evaluation and Review (CDER) utilizes a process referred to as the Voluntary Genomic Data Submission (VGDS) to allow for early and informal review of data generated during drug studies. A formal guidance has been developed to describe this process (http://www.fda.gov/cder/guidance/6400fnl.htm). Although the guidance does not address proteomics specifically, the principles of the guidance can be used as a model for proteomic submissions. These voluntary submissions are reviewed, not by a reviewing division, for example, the Division of Oncology Drug Products in CDER, but by a separate group, the Interdisciplinary Pharmacogenomic Review Group (IPRG) (http://www.fda.gov/cder/genomics/IPRG.htm). To date, VGDS are primarily submitted by pharmaceutical firms. If the firm is international in nature, and markets its products in Europe, the European Medicines Agency (EMEA) may also be asked to provide input as members of the IPRG (http://www.fda.gov/oia/ pilot-program0904.html). The guidance further defines the VGDS and the IPRG.

Included in the VGDS guidance is a discussion of analytical biomarkers. Three types of analytical biomarkers are described: known valid, probably valid, and exploratory biomarkers. Examples of known valid biomarkers include the enzyme biomarkers cytochrome P450 (2D6 and C19), thiopurine methyltransferase, and UGT1A1 *(15)*. This is a rapidly developing area and to date over two dozen informal submissions have been submitted with a broad variety of goals and with varying informal but beneficial educational and informational outcomes.

As part of a CDER outreach activity, the agency has developed a Cooperative Research & Development Agreement (CRADA) with Novartis Institutes for Biomedical Research. It is expected that the results of the CRADA, "Criteria for validation of genomic biomarkers of safety," will set criteria for the analytical validation of genomic biomarkers.

As part of CDRH outreach activity, the agency has sponsored over 100 visits by stakeholders with an area in molecular diagnostics including representatives of industry, academia, and other government agencies working in the areas of genomics, proteomics, and cancer biology. In addition, both CDER and CDRH have participated jointly in more than a half a dozen workshops on topics related to genomic test development and use.

5. SOME CRITICAL ISSUES WITH REGULATORY IMPLICATIONS FOR PROTEOMIC CANCER BIOMARKERS

The Human Genome Project is likely to provide many potential molecular targets with biologic plausibility for future drug development *(16)*. However, despite the research efforts and money being spent over the past decades and growing interest in biomarkers and their use in drug development, drug approval has not increased in recent years *(17)*.

Cancerous cells use many signaling pathways such as angiogenesis, proliferation, apoptosis, and metastasis with great biological robustness *(18)*. Cancer phenotype consists of a set of parameters of which the presentation is probably context dependent *(19)*. It is likely therefore that, for treatment success, multiple pathways may have to be targeted. However, delineating the drug targets at the molecular level and assigning physiological functions remain a formidable task. Moreover, validation of new biomarkers requires careful planning and consideration of all relevant scientific and regulatory issues. In this section, we will attempt to outline some important issues in the assay development of cancer proteomic biomarkers for oncology-targeted therapy. Because several articles have already been published on the regulatory perspective for biomarker development (*20–26*), we will be focusing only on some of the critical issues related to proteomic cancer biomarker assay development.

5.1. Target Characterization

Characterization of a new proteomic biomarker assay is best performed in the context of planned intended use *(27)*. A causal relationship between the potential target and outcome is also essential *(28)*. For example, it has been known that the type II topoisomerase is a possible drug target for anthracycline *(29)*. It has been identified as a putative target for anthracyclines *(30,31)*, which function as a topoisomerase poison. Topoisomerase IIα (TOP2A) gene aberrations have been detected in about 23% of the breast cancers by the fluorescence in situ hybridization method *(32)*. However, the gene amplification and deletion of TOP2A do not correlate well with the topoisomerase expression in tissue *(33)*. The increase of the enzyme in tissue detected by immunohistochemical (IHC) staining on breast cancer tissue section does not appear to change in parallel with the gene dosage. The topoisomerase IIα is affected by other factors such as p53 *(34)* and c-myb oncogene *(35)* that could modify the TOP2A gene expression. Therefore, the measurement of the target molecule at the DNA level does not seem to reflect the topoisomerase expression in the tissue *(36)*. It is uncertain whether testing of amplification of the TOP2A or topoisomerase itself will make for a better predictive marker for chemosensitivity of anthracycline. Similarly, HER2/neu is also thought to be a useful biomarker for patient triage and predicting treatment outcomes of the anthracycline-based adjuvant chemotherapy *(37)*. It is closely located on the same amplicon on chromosome 17q12-21 *(38)*. But in this case, the biological plausibility is less evident. In addition, there are also at least 22 other genes showing significant over-expression in breast cancer, including gain of chromosome 17q12-23 *(39)*, without clear-cut biological functions. In theory, they could be the potential drug targets of the anthracycline as well. Therefore, unequivocal causal correlation of an intended analyte as a potential drug target to outcomes and reliable measurements may be important factors in biomarker development.

It should be desirable to verify and validate the intended drug target by a second alternative approach. If an antibody is available, Western blots or IHC staining are valuable in validating molecular target identified by the mechanism-based drug discovery. To date, proteomics has played a limited role in both disease diagnosis and drug target discovery *(2)*. However, mass spectrometry is a promising technology to allow for direct tissue profiling *(40)*. The technology may be useful in future drug target validation especially for novel therapeutic targets for which antibody is not available. A drug target can be validated in another experimental system such as a cell-based model system at the transcription level or in an animal model at the systemic level *(41)*. In this case, the operational or case definition of the disease phenotype and measurability of the related target analytes and their relationship to the drug target should be clearly determined *(42,43)*. Recent development in short-interfering RNA (siRNA)-based techniques showed siRNA could be effective in modifying disease phenotypes *(2)* that could be useful for target characterization.

5.2. Intended Use and Indication for Use

Intended use describes how the device is to be used including information on the analyte to be measured. This includes the nature of the test (quantitative or qualitative, specimen type and matrix, as well as condition for use, i.e., prescription use or over the counter). Indications for use describe the condition or disease to be screened, diagnosed, monitored, or treated, the target patient population, and the frequency of use. FDA requests different types of data and statistical analyses in pre-market applications for commercial distribution of IVD devices. Specific claims for using the device must be supported by appropriate performance characteristics data. The type of data required depends on the intended use, indication for use, technological characteristics of the device, and on other claims made by the manufacturer.

In most of the drug development process, it is highly desirable to have assays to assess the efficacy and toxicity of the lead agents *(44–46)*. Ideally, the assays for early phase drug targets could be developed into clinical laboratory tests to triage patients and to predict treatment outcome. The predictive cancer biomarker could be the same drug target or could be derived from the molecules associated with the same or related signal transduction pathways *(47)*. The marker assays must be shown to be both analytically valid (e.g., have acceptable accuracy and reproducibility) for the drug developed and clinically valid (e.g., have acceptable diagnostic accuracy for the disease state or drug response of interest) for patient therapy choices. A laboratory test that will be used for a predictive marker to triage patients to receive specific therapy is unique in the sense that one needs to consider potential risks of the device to patients based on the proposed intended use and indication for use. The risks are gauged in terms of whether the testing can correctly identify the subset of the target patient population likely to respond to the target therapy, as well as how inaccurate test results (false-positive and false-negative results) will adversely affect patient outcomes. Other considerations include the extent of correlation of the cancer biomarker with clinically relevant patient endpoints and the balance between treatment benefit and potential drug toxicity.

The generalizability of the study results is an important characteristic indicative of external validity of a study *(48)*. It is important that the intended target patient population is reflected in the validation study. The study subjects should be representative of the future target patient population, and all subgroups of a target population should be included.

5.3. Method and Operation Standardization

The Human Proteomic Organization (HUPO) has initiated a standardization project on biomarker discovery and validation through the HUPO plasma proteome project *(49,50)*. It has laid the groundwork for future method standardization by identifying sources of variation of the mass spectrometry platforms for serum biomarker validation *(51)*. Method and operational standardization in proteomics on mass spectrometry platforms are critical in target validation especially in the pre-analytical phase of mass spectrometry measurement where activities such as sample handling and processing steps are prone to systemic bias *(52,53)*. It cannot be overemphasized that the laboratory operation of mass spectrometry should be standardized because the pre-analytical and analytical phases of testing have the potential to introduce significant systemic biases into mass spectrometry results.

Effective validation requires a priori acceptance criteria *(8)*. There are other important factors in analytical and post-analytical phases that could also significantly affect data quality of mass spectrometry. These are calibration, baseline subtraction or normalization, peak alignment, feature extraction and selection, quality controls, and quality assurance procedures for raw data acquisition and processing *(54–56)*. All steps in performing an assay under study should be standardized before initiating a study using appropriate reference materials, clearly written standard operating procedures and well-trained testing personnel. By building quality into the performance of studies, inter-laboratory, cross-platform, and inter-technologist variability can be minimized.

5.4. Machine Learning Algorithm

Proteomics studies of mass spectrometry will be compared to a reference ("gold") standard, most likely clinical diagnosis or patient outcomes. The initial (internal) evaluations will probably use banked specimens in a case control study using a supervised machine learning algorithm (57,58). In recent years, it has become generally accepted that cross-validation with an external and independent sample set is essential in the multiplex biomarker evaluation to prevent underestimation of false-positive and false-negative error rates (59,60).

Data variability can be significantly affected by the readout format of the mass spectrometry signals. For mass spectrometry methods that do not employ stable isotope as an internal standard, one usually measures peak height in determining signal intensity at a particular mass/charge ratio (m/z). It has been reported that the coefficient of variation of peak height signal intensity could be as high as 50–60% (61). Some researchers have therefore converted the peak altitude to a binary outcome to facilitate further data processing (61). However, such readout formats are not amenable to quantitative measurements. For this reason, stable isotopes are still used in quantitative mass spectrometry (62). The internal standard with stable isotope-labeled analog offers a relative measurement for the unknown concentration of a specific analyte. It also avoids the systemic biases due to ion suppression imposed by matrix effects (63) or abundance proteins (64). A new method to quantify protein and metabolites by mass spectrometry without isotopic labeling or spiked standards has been reported (65) that may lead to further method developments in biomarker discovery and validation. Post-instrumentation data processing, such as normalization, and peak alignment could also significantly change the data structure (52–55).

Some of the co-variables in multiplex testing may be derived from the same signaling pathway, thereby being highly correlated. The variables that are closely related in a multivariate model may be selectively reduced based on their underlying biology in facilitating further data processing and analysis (66). It is also important to examine the data distribution prior to selecting a machine learning algorithm that is appropriate for model fitting, and to check whether the model assumptions hold *(8)*.

It is uncertain at present how to biologically validate the machine learning algorithms, and what would be the appropriate quality control procedure for the derived proteomic multiplex testing. These are significant issues of both scientific and regulatory concern.

5.5. *Study Design*

5.5.1. INTERNAL AND EXTERNAL VALIDATION

Proteomic tumor markers should demonstrate that they are significant predictors of changing clinical status when used alone or in combination with other diagnostic modalities. This can be demonstrated by testing appropriate clinical samples in different development phases *(59,67,68)*. The model building and internal validation needed to demonstrate proof of concept may use a split sample design of a training set and a testing set with appropriate controls *(59)*. Validation is not expected to be as robust when using an internal testing set as when using an external testing set because the characteristics of the samples in the former will likely be very similar to those in the training set.

The validity of the cancer biomarker can be checked initially in an internal validation study as described above evaluating the accuracy of the marker in a case-control study. Ultimately, however, a randomized double-blinded prospective cohort study with independent sample sets, in conjunction with other known clinical or diagnostic co-variables such as age, gender, disease stage, remission, and recurrence *(68–72)*, may be needed to establish performance. In some cases, samples banked during prospective clinical trials can be used to replicate a prospective study. In review of data in this circumstance, attention needs to be paid to two key issues: stability of the analyte during storage and assurance that sample selection bias does not impact study results.

A standardized data gathering system and format for efficient clinical data collection including chart review, laboratory reports, pathology reports, and imaging modalities should be in place before the study begins *(73)*. Researchers are required to systematically document all protocols and provide data to support all diagnostic claims.

External validation should use an independent sample set *(58,59)*. The confounding effects of the multiple co-variables in an assay system should be controlled primarily in the study design stage *(74)*.

Randomization of the treatment assignment and testing order of the samples from different classes (e.g., diseased and non-diseased, drug responders, and non-responders) is probably the best approach to control multiple unknown co-variables and selection bias, which may confound data interpretation *(52,75,76)*. Statistical adjustment may be made but are not a substitute for a robust study design. The clinical trials on predictive markers will need a marker-based or marker by treatment interaction design *(76)*. Sample size calculation will need to assume higher event rates to achieve the same statistical power of a conventional clinical study when size of the treatment effect is evaluated alone *(70)*. The calculation made from studies should include strategies for dealing with missing data or sample attrition. Methods to compensate or correct for potential sampling bias and confounding factors are appropriate whenever such bias and confounding factors are unavoidable or becomes apparent a posteriori. The predictive marker may have to be approved concurrently with a targeted drug or together with the drug as a combination product or co-developed product *(25)*.

5.5.2. PATIENT OUTCOME AND SURROGATE ENDPOINTS

Clinical endpoints and patient outcome measures should be appropriate to the claims on indications of use. Primary patient outcomes for evaluating a predictive cancer biomarker are overall survival and progression-free survival *(77,78)*. A surrogate endpoint is a biomarker intended to substitute or predict the patient outcomes *(79,80)*. Development of criteria for statistical validation of surrogate endpoints continues to be a fervent area of research *(81–86)*. The research includes joint modeling of the clinical outcome (e.g., time to death or progression) with repeated measurements of the putative surrogate over time *(87)*. To be clinically useful in predicting patient outcomes, a candidate biomarker should demonstrate strong and consistent association with the natural history of the disease and the primary patient outcomes. It should be specific in tissue distribution, free from interference of matrix effects. Also its time course of concentration changes should occur in parallel with stage of disease progression and regression in exposing to therapeutics interventions in a dose-responsiveness manner *(79,80,88,89)*.

Objective clinical evidence of disease status such as symptoms and signs indicative of disease progression and regression, biopsy-derived histology, and/or imaging modalities could also be used as reference standards in validating tumor biomarker assays *(90–93)*. As new targeted agents are likely to be cytostatic rather than cytotoxic *(94)*, the objective or measurable treatment response may not be feasible in evaluation of targeted therapy. Under such circumstances, a composite reference standard may be constructed from clinical impression of physicians, imaging, and laboratory data *(93,95)*. A mechanism-based biological endpoint could also be used if its association can be demonstrated to be consistent with the patient outcomes *(96–99)*. A subjective, intangible, or multidimensional study endpoint such as health-related quality-of-life would be difficult to interpret *(92,95)*. Moreover, intermediate endpoints such as toxicity and treatment response may not be suitable in the evaluation of predictive markers *(91)*. A caution is that some of the intermediate outcome measures such as time-to-progression and time-to-recurrence or even tumor response may be difficult to interpret because their relationships to survival are largely undetermined *(92,96)*. Nonetheless, it is expected that new science-based biological surrogate endpoints could be useful in future clinical studies *(97–100)*.

The criteria for intermediate endpoints such as disease progression and treatment response should be specified or defined clearly prior to the study. For example, identification of the length of a favorable treatment response that can be counted as a meaningful disease remission or the extent of a mass shrinkage detected by imaging could be considered as an acceptable response to therapy.

5.5.3. Sample Size and Power

The study should be sufficiently powered to detect differences of desirable clinically significance effects. The sample size will have to be sufficiently large to accommodate all covariates for the study and sample attrition. The calculation of the sample size needs to factor in the biomarker prevalence *(101)*. It has been reported that as many as four times the number of events may be required for a marker by treatment study design to achieve the same statistical power in evaluating the same effect size as a regular intervention study *(70)*.

5.5.4. Statistical Modeling

The joint distribution of the biomarker in samples from healthy and disease subjects and the underlying assumptions of the statistical model are critical to the fitting of an appropriate statistical model to the data. It is important to examine joint distribution of the putative biomarkers and the assumptions underlying parametric models *(8,66)*. Although a distribution is not assumed for nonparametric methods, all other underlying assumptions of the multivariate model need to be met satisfactorily *(8)*.

In dealing with studies involving multiple endpoints, the Bonferroni correction can be used to adjust the probability of a false significance or Type I error (α) to control the chance of false-positive results at most α *(88,89,102)*. In the case of the highly correlated endpoints, the Bonferroni correction will be conservative *(103)*. In this case, p value resampling *(104)* or other methods *(60)* can be used to obtain a less severe yet still valid correction for multiplicity. Practical approaches using weighted parameters for developing a composite score based on biological evidence of the disease phenotype's modulation or other empirical methods may also be considered *(105,106)*. Also, the implication of a positive result for some particular analytes in the multivariate model to overall predictive power of the multiplex assay should be considered.

5.5.5. Supportive Evidence

Reproducibility of test results is critical for the validity of a laboratory test. For continuous valued tests, the coefficient of variation is useful for quantifying variation of the test result in sample replicates over days, operators, reagent lots, devices, and other conditions. For a high-dimensional laboratory test, a single study is insufficient to fully validate the clinical utility of the device. With proteomics data, the dimension of the candidate biomarker pool from which the test was developed often exceeds by many fold the number of samples studied. Consequently, over-fitting of the test results to the samples is an easy mistake to make, resulting in a substantial underestimation of test error rates. Cross-validation can be attempted to provide unbiased estimates of the error rates. But this can be done appropriately only if the machine learning algorithm for developing the test can be automated *(107)*. In reality, the process by which a test is developed is often not automated, but includes subjective biological considerations. Therefore, in most cases, a separate study would be needed for proper validation of a high-dimensional test.

Published literature using the same study device with similar design and data collection format could be admissible as part of supportive evidence for effectiveness and safety claim. A well-performed meta-analysis of all published clinical data pertinent to the analytes or device could be particularly useful in supplementing the validation data in a regulatory submission. When this is not possible, systemic reviews and other types of data analysis might be considered to support new biomarker submissions.

6. CONCLUSION

We have reviewed some of the relevant issues in cancer biomarker development and validation by proteomics for targeted cancer therapeutics. We have emphasized the importance of developing a clearly stated intended use and indication for use of the device, establishing a reliable laboratory measurement before initiating validation studies, addressing the need for biological validation of biomarkers resulting from machine learning algorithms, and using robust clinical study designs with representative patient populations and effective controls. It is clear that the regulation of devices based on large-scale protein science will continue to evolve and that many of the issues in regulation of predictive cancer biomarkers are currently inadequately defined. FDA is engaging all stakeholders in seeking input on determining appropriate validation criteria and a pre-market review process that is credible but meets the intent and spirit of The Food and Drug Administration Modernization Act, which calls on regulatory scientists to provide a "least burdensome" regulatory pathway and to focus on the critical regulatory thresholds involved in the pre-market review process. By working together, we hope to be partners rather then obstacles in the journey to transform the vast amount of the human genome data into patient-useful knowledge and to translate that knowledge into treatment benefits for cancer patients.

REFERENCES

1. Liu ET. Expression genomics and drug development: towards predictive pharmacology. Brief Funct Genomic Proteomic 2005 Feb;3(4):303–21.
2. Lindsey MA. Target discovery. Nat Rev Drug Discov 2003 Oct;2(10):831–8.
3. Chanda SK and Caldwell JS. Fulfilling the promise: drug discovery in the post-genomic era. Drug Discov Today 2003 Feb 15;8(4):168–74.
4. Alaiya A, Al-Mohanna, and Linder S. Clinical cancer proteomics: promises and pitfalls. J Proteome Res 2005 Jul–Aug;4(4):1213–22.
5. Walgren JL and Thompson DC. Application of proteomic technologies in the drug development process. Toxicol Lett 2004 Apr 1;149(1–3):377–85.
6. Pritchard KI, Shepherd LE, O'Malley FP, Andrulis IL, Tu D, Bramwell VH, and Levine MN, for the National Cancer Institute of Canada Clinical Trials Group. HER2 and responsiveness of breast cancer to adjuvant chemotherapy. N Engl J Med 2006 May 18;354(20):2103–11.
7. Feng Z, Prentice R, and Srivastava S. Research issues and strategies for genomic and proteomic biomarker discovery and validation: a statistical perspective. Pharmacogenomics 2004 Sep;5(6):709–19.
8. Mehta T, Tanik M, and Allison DB. Towards sound epistemological foundations of statistical methods for high-dimensional biology. Nat Genet 2004 Sep;36(9):943–7.
9. Hayes DF, Bast RC, Desch CE, Fritche H, Jr, Kemeny NE, Jessup M, Locker GY, Macdonald JS, Mennel RG, Norton L, Ravdin P, Taube S, and Winn RJ. Tumor

marker utility grading system: a framework to evaluate clinical utility of tumor markers. J Natl Cancer Inst 1996 Oct 16;88(20):1456–66.
10. McShane LM, Altman DG, Sauerbrei W, Taube SE, Gion M, Clark GM; Statistics Subcommittee of the NCI-EORTC Working Group on Cancer Diagnostics. Reporting recommendations for tumor marker prognostic studies (REMARK). J Natl Cancer Inst 2005 Aug 17;97(16):1180–4.
11. Schrorohl AS, Holten-Andersen M, Sweep F, Schmitt M, Harbeck N, Foekens J, and Brunner N, on behalf of the European Organization for Research and Treatment of Cancer (EORTC) Receptor and Biomarker Group. Tumor markers: from laboratory to clinical utility. Mol Cell Proteomics 2003 Jun;2(6):378–87.
12. Deyo RA and Jarvik JJ. New diagnostic tests: breakthrough approaches or expensive add-ons? Ann Intern Med 2003 Dec 2;139(11):950–1.
13. Bossuyt PM, Reitsma JB, Bruns DE, et al, for the STARD group. Towards complete and accurate reporting of studies of diagnostic accuracy: the STARD Initiative. Clin Chem 2003;49:1–6.
14. Schmitt M, Harbeck N, Daidone MG, Brunner N, Duffy MJ, Foekens JA, and Sweep FC. Identification, validation, and clinical implementation of tumor-associated biomarkers to improve therapy concepts, survival, and quality of life of cancer patients: tasks of the Receptor and Biomarker Group of the European Organization for Research and Treatment of Cancer. Int J Oncol 2004 Nov;25(5):1397–406.
15. Dervieux T, Meshkin B, and Neri B. Pharmacogenetic testing: proofs of principle and pharmacoeconomic implications. Mutat Res 2005 Jun 3;573(1–2):180–94.
16. Lindpaintner K. Pharmacogenetics and the future of medical practice. Br J Clin Pharmacol 2002 Aug;54(2):221–30.
17. Sams-Dodd F. Target-based drug discovery: is something wrong? Drug Discov Today 2005 Jan 15;10(2):139–47.
18. Vogelstein B and Kinzler KW. Cancer genes and the pathways they control. Nat Med 2004 Aug;10(8):789–99.
19. Segal E, Friedman N, Koller D, and Regev A. A module map showing conditional activity of expression modules in cancer. Nat Genet 2004 Oct;36(10):1090–8. Epub 2004 Sep 26.
20. Hackett JL and Gutmann S. Introduction to the Food and Drug Administration (FDA) regulatory process. J Proteome Res 2005 Jul–Aug;4(4):1110–3.
21. Bast RC, Jr, Lilja H, Urban N, Rimm DL, Fritche H, Grey J, Veltri R, Klee G, Allen A, Kim N, Gutman S, Rubin MA, and Hruszkewycz A. Translational cross-roads for biomarkers. Clin Cancer Res 2005 Sep 1;11(17):6103–8.
22. Gutman S. Regulatory issues in tumor marker development. Semin Oncol 2002 Jun;29(3):294–300.
23. Katz R. FDA: evidentiary standards for drug development and approval. NeuroRx 2004 Jul;1(3):307–16.
24. Katz R. Biomarkers and surrogate markers: an FDA perspective. NeuroRx 2004 Apr;1(2):189–95.
25. Harper CC, Phillip R, Robinowitz M, and Gutman SI. FDA perspectives on pharmacogenetic testing. Expert Rev Mol Diagn 2005 Sep;5(5):643–8.
26. Hirschfeld S and Pazdur R. Oncology drug development: United States Food and Drug Administration perspective. Crit Rev Oncol Hematol 2002 May;42(2):137–43.

27. Hammond ME and Taube SE. Issues and barriers to development of clinically useful tumor markers: a development pathway proposal. Semin Oncol 2002 Jun;29(3):213–21.
28. Rebbeck TR. Inherited genetic markers and cancer outcomes: personalized medicine in the postgenomic era. J Clin Oncol 2006, May1; 24(13):1972–4.
29. Piccart-Gebhart MJ. Anthracyclines and the tailoring of treatment for early breast cancer. N Engl J Med 2006 May 18;354(20):2177–9.
30. Walker JV and Nitiss JL. DNA topoisomerase II as a target for cancer chemotherapy. Cancer Invest 2002;20(4):570–89.
31. Nitiss JL and Beck WT. Antitopoisomerase drug action and resistance. Eur J Cancer 1996 Jun;32A(6):958–66.
32. Knoop AS, Knudsen H, Balslev E, Rasmussen BB, Overgaard J, Nielsen KV, Schonau A, Gunnarsdottir K, Olsen KE, Mouridsen H, and Ejlertsen B. Retrospective analysis of topoisomerase IIa amplifications and deletions as predictive markers in primary breast cancer patients randomly assigned to cyclophosphamide, methotrexate, and fluorouracil or cyclophosphamide, epirubicin, and fluorouracil: Danish Breast Cancer Cooperative Group. J Clin Oncol 2005 Oct 20;23(30):7483–90.
33. Mueller R, Parkes RK, Andrulis I, and O'Malley FP. Amplification of the TOP2A gene does not predict high levels of topoisomerase II alpha protein in human breast tumor samples. Genes Chromosomes Cancer 2004 Apr;39(4):288–97.
34. Sandri MI, Isaacs RJ, Ongkeko WM, Harris AL, Hickson ID, Broggini M, and Vikhanskaya F. p53 regulates the minimal promoter of the human topoisomerase IIalpha gene. Nucleic Acids Res 1996 Nov 15;24(22):4464–70.
35. Brandt TL, Fraser DJ, Leal S, Halandras PM, Kroll AR, and Kroll DJ. c-Myb trans-activates the human DNA topoisomerase IIalpha gene promoter. J Biol Chem 1997 Mar 7;272(10):6278–84.
36. Ross JS, Fletcher JA, Bloom KJ, Linette GP, Stec J, Symmans WF, Pusztai L, and Hortobagyi GN. Targeted therapy in breast cancer: the HER-2/neu gene and protein. Mol Cell Proteomics 2004 Apr;3(4):379–98.
37. Di Leo A, Gancberg D, Larsimont D, Tanner M, Jarvinen T, Rouas G, Dolci S, Leroy JY, Paesmans M, Isola J, and Piccart MJ. HER-2 amplification and topoisomerase IIalpha gene aberrations as predictive markers in node-positive breast cancer patients randomly treated either with an anthracycline-based therapy or with cyclophosphamide, methotrexate, and 5-fluorouracil. Clin Cancer Res 2002 May;8(5):1107–16.
38. Jarvinen TA and Liu E. HER-2/neu and topoisomerase IIalpha in breast cancer. Breast Cancer Res Treat 2003 Apr;78(3):299–311.
39. Willis S, Hutchins AM, Hammet F, Ciciulla J, Soo WK, White D, van der Spek P, Henderson MA, Gish K, Venter DJ, and Armes JE. Detailed gene copy number and RNA expression analysis of the 17q12–23 region in primary breast cancers. Genes Chromosomes Cancer 2003 Apr;36(4):382–92.
40. Chaurand P, Sanders ME, Jensen RA, and Caprioli RM. Proteomics in diagnostic pathology: profiling and imaging proteins directly in tissue sections. Am J Pathol 2004 Oct;165(4):1057–68.
41. Lindsey MA. Finding new drug targets in the 21st century. Drug Discov Today 2005 Dec;10(23–24):1683–7.
42. Strohman R. Maneuvering in the complex path from genotype to phenotype. Science 2002 Apr 26;296(5568):701–3.

43. Coggon D, Martyn C, Palmer KT, and Evanoff B. Assessing case definitions in the absence of a diagnostic gold standard. Int J Epidemiol 2005 Aug;34(4): 949–52.
44. Colburn WA and Lee JW. Biomarkers, validation and pharmacokinetic-pharmacodynamic modelling. Clin Pharmacokinet 2003;42(12):997–1022.
45. Colburn WA. Biomarkers in drug discovery and development: from target identification through drug marketing. J Clin Pharmacol 2003 Apr;43(4):329–41.
46. Frank R and Hargreaves R. Clinical biomarkers in drug discovery and development. Nat Rev Drug Discov 2003 Jul;2(7):566–80.
47. Fishman MC and Porter JA. Pharmaceuticals: a new grammar for drug discovery. Nature 2005 Sep 22;437(7058):491–3.
48. Wells KB. Treatment research at the crossroads: the scientific interface of clinical trials and effectiveness research. Am J Psychiatry 1999 Jan;156(1):5–10.
49. Semmes OJ, Feng Z, Adam BL Banez LL, Bigbee WL, Campos D, Cazares LH, Chan DW, Grizzle WE, Izbicka E, Kagan J, Malik G, McLerran D, Moul JW, Partin A, Prasanna P, Rosenweig J, Sokoll LJ, Srivastava S, Srivastava S, Thompson I, Welch MJ, White N, Winget M, Yasui Y, Zhang Z, and Zhu L. Evaluation of serum protein profiling by surface-enhanced laser desorption/ionization time-of-flight mass spectrometry for the detection of prostate cancer: I. Assessment of platform reproducibility. Clin Chem 2005 Jan;51(1):102–12.
50. Kapp EA, Schutz F, Connolly LM, Chakel JA, Meza JE, Miller CA, Fenyo D, Eng JK, Adkins JN, Omenn GS, and Simpson RJ. An evaluation, comparison, and accurate benchmarking of several publicly available MS/MS search algorithms: sensitivity and specificity analysis. Proteomics 2005 Aug;5(13):3475–90.
51. Omenn GS. Advancement of biomarker discovery and validation through the HUPO plasma proteome project. Dis Markers 2004;20(3):131–4.
52. White CN, Chan DW, and Zhang Z. Bioinformatics strategies for proteomic profiling. Clin Biochem 2004 Jul;37(7):636–41.
53. Villanueva J, Phillip J, Chaparro CA, Li Y, Toledo-Crow R, deNoyer L, Fleisher M, Robbins R, and Tempst P. Correcting common errors in identifying cancer-specific serum peptide signatures. J Proteome Res 2005 Jul–Aug;4(4):1060–72.
54. Shin H and Markey MK. A machine learning perspective on the development of clinical decision support systems utilizing mass spectra of blood samples. J Biomed Inform 2006 Apr;39(2):227–48.
55. Listgarten J and Emili A. Statistical and computational methods for comparative proteomic profiling using liquid chromatography-tandem mass spectrometry. Mol Cell Proteomics 2005 Apr;4(4):419–34.
56. Lindon JC, Nicholson JK, Holmes E, et al, for The Standard Metabolic Reporting Structure Working Group. Summary recommendations for standardization and reporting of metabolic analyses. Nat Biotechnol 2005 Jul;23(7):833–8.
57. Zhang Z and Chan DW. Cancer proteomics: in pursuit of "true" biomarker discovery. Cancer Epidemiol Biomarkers Prev 2005 Oct;14(10):2283–6.
58. Simon R. Development and validation of therapeutically relevant multi-gene biomarker classifiers. J Natl Cancer Inst 2005 Jun 15;97(12):866–7.
59. Simon R. Roadmap for developing and validating therapeutically relevant genomic classifiers. J Clin Oncol 2005 Oct 10;23(29):7332–41.
60. Allison DB, Cui X, Page GP, and Sabripour M. Microarray data analysis: from disarray to consolidation and consensus. Nat Rev Genet 2006 Jan;7(1):55–65.
61. Yasui Y, Pepe M, Thompson ML, Adam BL, Wright GL, Jr, Qu Y, Potter JD, Winget M, Thornquist M, and Feng Z. A data-analytic strategy for protein

biomarker discovery: profiling of high-dimensional proteomic data for cancer detection. Biostatistics 2003 Jul;4(3):449–63.
62. Tao WA and Asbersold R. Advances in quantitative proteomics via stable isotope tagging and mass spectrometry. Curr Opin Biotechnol 2003 Feb;14(1):110–8.
63. Schlosser G, Pocsfalvi G, Huszar E, Malorni, and Hudecz F. MALDI-TOF mass spectrometry of a combinatorial peptide library: effect of matrix composition on signal suppression. J Mass Spectrom 2005 Dec;40(12):1590–4.
64. Sun W, Wu S, Wang X, Zheng D, and Gao Y. An analysis of protein abundance suppression in data dependent liquid chromatography and tandem mass spectrometry with tryptic peptide mixtures of five known proteins. Eur J Mass Spectrom (Chichester, Eng) 2005;11(6):575–80.
65. Wang W, Zhou H, Lin H, Roy S, Shaler TA, Hill LR, Norton S, Kumar P, Anderle M, and Becker CH. Quantification of proteins and metabolites by mass spectrometry without isotopic labeling or spiked standards. Anal Chem 2003 Sep 15;75(18):4818–26.
66. Feng Z and Yasui Y. Statistical considerations in combining biomarkers for disease classification. Dis Markers 2004;20(2):45–51.
67. Zolg JW and Langen H. How industry is approaching the search for new diagnostic markers and biomarkers. Mol Cell Proteomics 2004 Apr;3(4):345–54.
68. Pepe MS, Etzioni R, Feng Z, Potter JD, Thompson ML, Thornquist M, Winget M, and Yasui Y. Phases of biomarker development for early detection of cancer. J Natl Cancer Inst 2001 Jul 18;93(14):1054–61.
69. Sledge GW, Jr. What is targeted therapy? J Clin Oncol 2005 Mar 10;23(8): 1614–5.
70. Paik S. Clinical trial methods to discover and validate predictive markers for treatment response in cancer. Biotechnol Annu Rev 2003;9:259–67.
71. McShane LM, Altman DG, and Sauerbrei W. Identification of clinically useful cancer prognostic factors: what are we missing? J Natl Cancer Inst 2005 Jul 20;97(14):1023–5.
72. Conley BA and Taube SE. Prognostic and predictive markers in cancer. Dis Markers 2004;20(2):35–43.
73. Jansen AC, van Aalst-Cohen ES, Hutten BA, Buller HR, Kastelein JJ, and Prins MH. Guidelines were developed for data collection from medical records for use in retrospective analyses. J Clin Epidemiol 2005 Mar;58(3):269–74.
74. Colantonio DA and Chan DW. The clinical application of proteomics. Clin Chim Acta 2005 Jul 24;357(2):151–8.
75. Hu J, Coombes KR, Morris JS, and Baggerly KA. The importance of experimental design in proteomic mass spectrometry experiments: some cautionary tales. Brief Funct Genomic Proteomic 2005 Feb;3(4):322–31.
76. Sargent DJ, Conley BA, Allegra C, and Collette L. Clinical trial designs for predictive marker validation in cancer treatment trials. J Clin Oncol 2005 Mar 20;23(9):2020–7.
77. American Society of clinical Oncology. Outcomes of cancer treatment for technology assessment and cancer treatment guidelines. American Society of Clinical Oncology. J Clin Oncol 1996 Feb;14(2):671–9.
78. Hirschfeld S and Pazdur R. Oncology drug development: United States Food and Drug Administration perspective. Crit Rev Oncol Hematol 2002 May;42(2): 137–43.
79. Colburn WA. Optimizing the use of biomarkers, surrogate endpoints, and clinical endpoints for more efficient drug development. J Clin Pharmacol 2000 Dec;40(12 Pt 2):1419–27.

80. Floyd E and McShane TM. Development and use of biomarkers in oncology drug development. Toxicol Pathol 2004 Mar–Apr;32 Suppl 1:106–15.
81. Prentice RL. Surrogate endpoints in clinical trials: definition and operational criteria. Stat Med 1989 Apr;8(4):431–40.
82. Fleming TR and DeMets DL. Surrogate end points in clinical trials: are we being misled? Ann Intern Med 1996 Oct 1;125(7):605–13.
83. Buyse M and Molenberghs G. Criteria for the validation of surrogate endpoints in randomized experiments. Biometrics 1998 Sep;54(3):1014–29.
84. Buyse M, Molenberghs G, Burzykowski T, Renard D, and Geys H. The validation of surrogate endpoints in meta-analyses of randomized experiments. Biostatistics 2000 Mar;1(1):49–67.
85. Burzykowski T, Molenberghs G, Buyse M, Geys H, and Renard D. Validation of surrogate endpoints in multiple randomized clinical trials with failure time end points. Appl Stat 2001 50:405–22.
86. DeGruttola VG, Clax P, DeMets DL, Downing GJ, Ellenberg SS, Firedman L, Gail M, Prentice R, Wittes J, and Zeger S. Consideration in the evaluation of surrogate endpoints in clinical trials: summary of a National Institutes of Health Workshop. Control Clin Trials 2001, 22:485–502.
87. Xu J, Zeger SL. Joint analysis of longitudinal data comprising repeated measures and times to events. Appl Stat 2001 50:375–387.
88. Mayer D. Essential Evidence-Based Medicine. Cambridge University Press, UK, USA, Australia, Spain, South America; Bk&CD-Rom edition (June 17, 2004).
89. Fletcher RH and Fletcher SW. Clinical Epidemiology: The Essentials. Lippincott Williams & Wilkins, Philadelphia, Baltimore, New York, London, Buenos Aires Hong Kong, Sidney, Tokyo; 4th edition (March 2005)
90. Lesko LJ and Atkinson AJ, Jr. Use of biomarkers and surrogate endpoints in drug development and regulatory decision making: criteria, validation, strategies. Annu Rev Pharmacol Toxicol 2001;41:347–66.
91. Gelmon KA, Eisenhauer EA, Harris AL, Ratain MJ, and Workman P. Anticancer agents targeting signaling molecules and cancer cell environment: challenges for drug development? J Natl Cancer Inst 1999 Aug 4;91(15):1281–7.
92. Schilsky RL. End points in cancer clinical trials and the drug approval process. Clin Cancer Res 2002 Apr;8(4):935–8.
93. Williams G, Pazdur R, and Temple R. Assessing tumor-related signs and symptoms to support cancer drug approval. J Biopharm Stat 2004 Feb;14(1): 5–21.
94. Fox E, Curt GA, Balis FM. Clinical trial design for target-based therapy. Oncologist 2002;7(5):401–9.
95. Johnson JR, Williams G, and Pazdur R. End points and United States Food and Drug Administration approval of oncology drugs. J Clin Oncol 2003 Apr 1;21(7):1404–11.
96. Meyerson LJ, Wiens B, LaVange LM, and Koutsoukos AD. Quality control of oncology clinical trials. Hematol Oncol Clin North Am 2000 Aug;14(4):953–71, x.
97. Cooper R and Kaanders JH. Biological surrogate end-points in cancer trials: potential uses, benefits and pitfalls. Eur J Cancer 2005 Jun;41(9):1261–6.
98. Kelloff GJ and Sigman CC. New science-based endpoints to accelerate oncology drug development. Eur J Cancer 2005 Mar;41(4):491–501.
99. Korn EL, Arbuck SG, Pluda JM, Simon R, Kaplan RS, and Christian MC. Clinical trial designs for cytostatic agents: are new approaches needed? J Clin Oncol 2001 Jan 1;19(1):265–72.

100. Park JW, Kerbel RS, Kelloff GJ, Barrett JC, Chabner BA, Parkinson DR, Peck J, Ruddon RW, Sigman CC, and Slamon DJ. Rationale for biomarkers and surrogate end points in mechanism-driven oncology drug development. Clin Cancer Res 2004 Jun 1;10(11):3885–96.
101. Sargent D and Allegra C. Issues in clinical trial design for tumor marker studies. Semin Oncol 2002 Jun;29(3):222–30.
102. Pajak TF, Clark GM, Sargent DJ, McShane LM, and Hammond ME. Statistical issues in tumor marker studies. Arch Pathol Lab Med 2000 Jul;124(7):1011–5.
103. Lagakos SW. The challenge of subgroup analyses—Reporting without distorting. N Engl J Med 2006 April 20, 354(16):1667–9.
104. Westfall PH and Young SS. Resampling-Based Multiple Testing: Examples and Methods for p-Value Adjustment. Wiley Series in Probability and Statistics, New York, Canada (1993).
105. Ramnarayan P, Kapoor RR, Coren M, Nanduri V, Tomlinson AL, Taylor PM, Waytt JC, and Brotto JF. Measuring the impact of diagnostic decision support on the quality of clinical decision making: development of a reliable and valid composite score. J Am Med Inform Assoc 2003 Nov–Dec;10(6):563–72.
106. Hwang D, Rust AG, Ramsey S, Smith JJ, Leslie DM, Weston AD, de Atauri P, Aitchison JD, Hood L, Siegel AF, and Bolouri H. A data integration methodology for systems biology. Proc Natl Acad Sci USA 2005 Nov 29;102(48):17296–301.
107. Simon RM, Korn EL, McShane LM, Radmacher MD, Wright GW, and Zhao Y. Design and Analysis of DNA Microarray Investigations. Springer, New York (2003).

INDEX

Absolute quantification, method for, 165
Accentuation of differentially expressed proteins using phage technology, 90
Accurate mass tag methods, 11
Acute lymphoblastic leukemia 162, 172
Acute myeloid leukemia 161–2, 169, 172
   cell extracts in classification, 170
   cytogenetic diagnostics (karyotyping) of, 169
   diagnostic risk stratification of, 181
Adenocarcinoma, 140
Adenomatous hyperplasia, 143
Affimetrix microarray techniques, 124
Affinity chromatography, 21, 179
   introduction of, 168
affinity tagging, 220
Ak mouse thymoma, 107
All-trans retinoic acid 163, 174–5
Alpha fetoprotein, 103, 191, 202, 242
Alternative splicing database, 215
Amino acid sequences, 102
AML *see* Acute myeloid leukemia
Analytical sensitivity, 196
Anaplastic lymphoma kinase, 70
Antibody microarrays, 84–5
Anticancer drug effects, 192
Anti-cancer therapy, 199
Autoantibodies, identification of, 132
Automaticgain control measures, 6
Avidin chromatography, 219

Bayer diagnostics, 94
Bence–Jones protein, 191
Bevacizumab, 104, 114
Biological fluids, 126
Biomarker, 64, 68, 81–5, 93, 243, 248
   degradation, 194
   development, 102
   discovery, 103, 111,
   advances in discovery, 129–31
   complexity of, 112
   for oncology-targeted therapy, 244
   biological validation of, 252

   discovery of, 98
   eTag assay system for, 87
   for the early detection of cancer, 198–9
Biopsy and array paradigm, 114
Bladder cancer, 103, 109
Blood markers, 192
Bone-forming cells, 176
Bonferroni correction, 251
Breast cancer, 60, 68, 71, 94, 95, 103, 105, 173, 179, 191, 198–201, 244
   development of first humanized antibody, 66
   endocrine therapy in, 191
   Oncoprotein as biomarkers for, 94–5
Bronchoalveolar lavage fluid, 148

Calibrator, 195
Cancer Cell Mitochondria, proteomic analysis of, 90–1
Cancer
   therapy, proteins as targets for, 91
   cancer tissue proteomics, 91–2
   management of, 82
   molecular diagnostics of, 88
Carboxylation, 18
Carcino embryonic antigen, 103, 191, 242
Carcinoma tissues, 142
Cardiac hypertrophy, 102
CDDP chemotherapy, 153
Cell population heterogeneity, 169
Cell proliferation, 175, 217
Cellular stresses, 41
Chemosensitivity of anthracycline, 244
Chemotherapy, 39, 60, 66, 72, 88, 92, 95–6, 113, 122, 140, 153, 163, 17–74, 200–1,
ChIPmicroarrays, 42
Chromatin immunoprecipitation, 41, 43–5
Chromatography, 7, 149
Chromosomal alterations, 123
   miscellanea of, 129
Chronic lymphocytic leukemia, 167
chronic myelogenous leukemia, 60
Chronic myeloid leukemia, 163

259

Circulating nucleosomes, detection of, 88
Cisplatin-based chemotherapy, 153
Class prediction model, 145
Classical proteomics strategies, limitations of the, 130
Clinical phosphoproteomics, 71–2
Clinical sensitivity, 199
Clinical trial, 60, 62, 66, 72, 93, 96, 101–2, 106, 108, 111–5, 176, 178, 191–2, 200–4, 248
  molecular proteomic techniques for, 104–11
ClinProt technique, 126, 131
Cohybridization, 42
Co-immunoprecipitation, 52
Collisional cooling, absence of, 12
Collision-induced dissociation, 6
Colon cancer, 40, 90, 112, 178, 180, 199–200
Colorectal cancer, 47, 64, 67, 72, 103, 191
Combination Therapy, 114
Combinatorial therapy, 104
Cox modeling, 155
Cysteine-specific labels, 27
Cytochemistry, 169
Cytogenetics and molecular diagnostics, 161
Cytokeratins, 144
Cytological examination, 143, 169,

Data dissemination, 221–3
Deamidation, 18
DeCyder dedicated software, 130
Diabetic nephropathy/retinopathy, 102
Diagnostic biomarkers, 170
Diesel exhaust particles, 153
Disease-related volatile organic compounds, 139
Disintegration/extraction of tissue, 194
DNA
  arrays, 167
  coding sequences, identification of, 125
  damage, 45–6, 49
    mitotic spindle, 41
  replication, 217
  sequences, 39, 225
Double blind trial, 96
Drug
  related molecular pathology, 192
  resistance, 180
  toxicity, biological mechanisms of, 174
Dysplasia, 139, 142

Edman sequencing technique, 225
Electron capture dissociation 8, 16–9
Electron transfer dissociation 19–20
Electronic noses, 149
Electrospray ionizationquadrupole-time-of-flight, 5
ELISA test, *see* Enzyme-linked immunoabsorbant assay
Endoplasmic Reticulum, 216
Enzyme-linked immunoabsorbant assay 24, 48, 94, 131, 155, 166
Epidermal growth factor, 26, 62, 87
Epitope, 127, 166, 196
Eponomycin, 179
Epoxomicin, 179
Epsilon amino groups, 129
Erlotinib, 96
Exhaled air, 149–51
External quality assessment, 197

Fecal occult blood testing, 199
Femtomolar sensitivities, 10
Fibroblast growth factor, 63
Fluoresence in situ hybridization, 191
Food and Drug Administration, 103, 241, 245, 252
Food and Drug Administration Modernization Act, 252
Fourier transform-ion cyclotron resonance 5, 9–12
Fragmentation pattern, 16
Functional proteomics studies, 151–5

Gastrointestinal stromal tumors 66
Gefitinib, 60, 66, 95, 96, 104, 113–4
Gel electrophoresis, 105–6
GeLC experiments, 20
Gel-free methods for protein expression, 219
Gene expression
  analysis, 97, 162
  map, 42
  studies, 163
Gene function, systematic interrogation of, 211
Gene modifications, 129
Gene signatures, 200
Genes and proteins, measurement of, 40
Genome-scale interaction, 220
Genomic amplification, 102
Genomic Contribution, 123–5

Index

Gleevec, 60, 66–7, 69, 83, 175
   discovery of, 65
Glycoprotein-imprinted gels, 88
Glycosylation, 18, 216, 226
Gradient polyacrylamide gel, 146
Green fluoresce protein, 217
GR-Golgi intermediate compartment, 218
Growth hormone disorders, 102

Helix-loop-helix transcription factors, 97
Herceptin, 66, 83
Heterogeneous nuclear ribonucleoprotein, 143
High-field asymmetric waveform ion mobility mass s, 93
High-pressure liquid chromatography, 164
High-throughput Analyses, 106–11
"high-risk" patients, 199–200
Histocompatibility complex, 124
Histone deacetylase inhibitors, 177–8
Histopathology, 129
Hormone-activated nuclear receptor, 53
Human chorionic gonadotrophin, 202
Human epidermal growth factor-2, measurement of, 201
Human genome project, 164, 225, 243
Human plasma proteome project, 156, 246
Human protein atlas, 165
Human proteome initiative, 225
Human proteomic organization, 246
Hybrid mass spectrometer system, 12
Hypermethylation, 125
Hypoxia, 41
   induced gene family, 127

ICAT *see* isotope-coded affinity tag
Immobilized metal affinity chromatography, 21–2
Immunoaffinity purification, 167
Immunoblot analysis, 105–6
Immunohistochemical studies, 114, 244
Immunohistochemistry, 95, 105, 167, 218
Immunophenotypic analyses, 169
Immunoprecipitated proteins, 50
Immunotherapy, 122
In vitro diagnostic device, 240–1, 245
Interdisciplinary pharmacogenomic review group, 242
Intraassay precision performance, 195
Ion activation methods, 30
Ion fragmentation methods, 15–20
Ion trapping region, 8

Isotope-coded affinity tag (ICAT) 20, 217
Isotope-coding, 26

Karyotyping, 169

Label-free methods, 27
Labeling methods, 24
Labelling by biotin reagents, 126
Lactacystin, 179
Lactate dehydrogenase, 127
Laser capture microdissection, 85
   chromatograms, 29
Leukemia, graft-versus-host proteomics in, 180–1
Leukemogenic process, 67
Linear ion traps, 5–9
Linearity, 196
Liquid chromatography, 7, 10
   automation of, 13
LIT instrument, 8
Liver cancer, 103
"low-risk" patients, 199
Lung adenocarcinoma, 144
Lung cancer, 140, 146, 152
   biomarkers of, 149
   cell lines, 144–5
   classification, 141
   molecular classification of, 140
   non-small cell, 66
   personalized management of, 95–6
   proteomics in, 139
Lung carcinoma, histologic types of, 140
Lung tumors, Morphologic assessment of, 155

Mabthera, 83
Machine learning algorithm, 247
MALDI mass spectrum, 52, 92
MALDI-Tandem MS, 13–5
Malignant tissue, 71
Mass analyzer design, 3
Mass spectrometry, 3–31, 85, 106, 108–11, 142, 217
   instrumentation, 5–6
   pre-analytical phase of, 131, 246
   technology, 164
Matrix-assisted laser desorption/ionization-time-o, 5
Medical device amendments, 240
Metabolic changes, 41
Metalloproteinase receptors, 143

Metaplasia, 139, 142
Methylation-based markers, 8, 125
Microarray
  based transcript profiling, 42–3
  hybridization, 46
  technology, 123
Mitotic spindle, 41, 218
Molecular interaction mapping, 40
Monogram biosciences, 87
Monospecific antibodies, 218
Morphoproteomics, 92
Mudpit proteomic experiments, 7
Multi-center clinical trial, 102, 112
Multidimensional protein identification technology, 70
Multiplexed protein assays, 165–7
Multistep activation schemes, 18
Mutation assay, 96
Mutation-positive women, 109
Myoglobin, 29

Nanobiotechnologies, 98
Nephron-sparing surgery, 122
Node-negative patients, 200
Noninvasive tumor-specific molecular imaging, 86
Non-malignant cells, 194
Non-synonymous mutations, 67
Novel Cellular Processes, Discovery of, 154
Nuclear Magnetic Resonance (NMR) 24, 216
Nuclear matrix protein, 103

'omics' technologies, 212, 228
Oncogene activation, 41
Oncology drug
  CellCarta technology for, 89–90
  diagnostics, 237
  products, division of, 242
  proteomics in, 89–93
Oncoproteomics, 81–2, 92
  application of nanobiotechnology in, 92–3
  development of, 98
Orbitrap, 6, 9, 12–3
Ovarian cancer, 102–3, 106, 108–10, 112–4, 191, 203–4

P53
Molecular
  sequencing, 84
  signaling

genomics of, 41–6
proteomics in, 49–54
target identification 47
Pancreatic cancer, 103
Partial nephrectomy, 122
Patient-tailored molecular medicine, emerging era of, 72
PCR gene expression analysis, 130
Peptide mass fingerprint, 164
Peptide-labeling methods, 30
Personalized medicine, 82
Personalized treatment of cancer, 96
  development of, 82
PET sequencing approach, 45
Phage display technology, 90
Philadelphia chromosome, 65
Phosphoproteomics, 21–3
  analysis, 68, 70
Phosphorylation, 23, 62, 107, 142, 215
  sites, 8
    localization of, 17
  status, 72
Phosphotyrosyl binding (PTB) domains, 63
PKC phosphorylation, 50
Platelet derived growth factor receptor (PDGFr) 26, 175, 201
Pleural effusion, 147–8
Post-genome research, 228
Postoperative surveillance, 202
Post-translational modification 126, 214–7, 219
Pre-analytical phase, 131
Pre-mammography, 198
Pre-market approval, 240
Primary colorectal, differential characteristics of, 107
Prostate cancer, 83–6, 103, 109, 112, 191, 198
  biology of the, 102
  development of, 97
  personalized management of 96–8
Prostate-specific antigen 83, 103, 198, 242
Prostatic hypertrophy, 109
Proteasome inhibitors, 178–9
Protein
  analysis methods, 20–31
  and peptides, separation of, 164
  arrays, 106
  biomarkers, analysis of, 84
  data bank, 225
  kinases, genomic and structural features of, 65–7

Index

microarrays, 71, 107
profiling approaches, 167
–protein interaction, 220–1
proteolysis of, 215
tyrosine kinases 60
tyrosine phosphatases, 64
ProteinChip system, 147
Proteolytic processing, 126, 216
Proteomic
   advances in, 4
   based clinical trials, 102
   biomarker assay, 244
   data, global advantages of, 223
   in discovery of biomarkers, 83–8
   of body fluids, 146
   of brain cancer, 98
   profiling, 70, 131
   role in clinical trials, 93–4
   role in management cancers, 94
   stratification, 214
   technologies, 82, 98, 112
   tumor markers, 248

Q-TOF instrument, 7
Q-Trap instrument, 8
Quadrupole ion traps, 6
Quantitative techniques, 23–30

Radical nephrectomy, 122
Receptor phosphorylation, 63–4
Redox system, 152–3
Renal cell carcinoma, 121–2
Responder profiling, 153–4
Retinoic acid receptor, 163
RNA interference, 46–7

Saccharomyces cerevisiae genome, 46
Screening for cervix cancer, 112
SELDI technology, 132, 147
Sequencing approach, 44
Serial analysis of binding elements, 45–6
Serial analysis of gene expression (SAGE) 42, 46, 48
Serological
   diagnosis, 154
   tumor biological marker, 198
Serum, 146–7
   proteins, Analysis of, 147
   proteomics, 101–2
      Advances in, 115

samples, 113
Short-interfering RNA, 245
Shotgun sequencing, 20
Silver staining, 148
Single-agent therapy, 114
Single-cell analysis of signaling responses, 173–4
Single-cell flow cytometric analysis, 173
Solid-phase microextraction, 149
Somatic mutations, 95
Sorafenib, 104, 114
Splice variants, 214–5
Squamous carcinoma, 142–3
Stable isotope labeling, 26, 165
Standard statistical procedures, 197
Statistical modeling, 250
Strong cation exchange, 21
Sub-cellular fractionation methods, 86, 217
Sub-cellular localization, 217
Sumylation, 18
Surface-enhanced laser desorption 85, 108, 126, 139, 143, 165
Surrogate markers, 64, 93

Tandem affinity purification, 70
Taxol-induced apoptosis, 71
Techniques and instrumentation for clinical proteomics, 110
Telomerase activity, measurement of, 84
TGF-β pathway, 43
Therapeutic interventions, 92
Therapeutic synergism, 104
Therapy predictive biomarker, 190
Thymidine phosphorylase, 127–8
Tissue
   and cell lines, analysis of, 127
   expression, 218–9
   heterogeneity, 72
   proteomics, 86–7
   sampling, 72
Top-down sequencing, 15, 30–1
Toxicology studies, 153
Transcriptomics technologies, 97, 142
Transplantation rejection, 102
Treatment trials, 111, 113–4
Triosephosphate isomerase, 144
Tubulin polymerization assay, 95
Tumor biomarkers, 189–205
   classification of, 242
   detection of, 86, 88

examples of, 191
monitoring, 202–4
utility grading system, 191
Tumor biopsies, 113
Tumor cell
 cultures, analysis of, 91
 lines, 127
Tumor correlated proteins, 126
Tumor of unknown primary, 92
Tumor-associated antigens, identification of, 154–5
Two-dimensional gel electrophoresis, 68, 85, 98, 126, 139, 142, 151, 153, 155, 215, 217, 219
 advantages of, 85
Two-dimensional Western blot analysis expression, 151
Tyrosine kinase, 62–4
 circuitry, 71
 inhibitors, 67, 175–7
 proteomic profiling of, 68–72
 receptor activation, 62–3
 signaling diversity, 64

Ubiquitin thiolesterase, 144
Ultramicroarrays, 84
Undifferentiated small cell lung carcinoma, 140
Union International Contre le Cancer (UICC) 123
UniProt knowledgebase, 217, 223–5
Urokinase plasminogen activator, 200

Vascular endothelial growth factor, 60, 154
Voluntary Genomic Data Submission, 242

Western blots, 154, 167
WHO disease classification, 162, 169

Zeptomolar sensitivities, 10

Printed in The United States of America